# TOMATOES

## The Multi-Plant Method

LEOPOLD KLEIN

# TOMA

# The Multi-P

# TOES

## lant Method

PRENTICE-HALL, INC., Englewood Cliffs, N.J.

**TOMATOES The Multi-Plant Method**

***by Leopold Klein***


Printed in the United States of America
Prentice-Hall International, Inc., London
Prentice-Hall of Australia, Pty. Ltd., Sydney
Prentice-Hall of Canada, Ltd., Toronto
Prentice-Hall of India Private Ltd., New Delhi
Prentice-Hall of Japan, Inc., Tokyo
10 9 8 7 6 5 4 3 2 1

Library of Congress Cataloging in Publication Data
Klein, Leopold.
Tomatoes.
Includes index.
1. Tomatoes. I. Title.
SB349.K44 1976 635'.642 76-3694
ISBN 0-13-924761-0

*Design by Linda Huber*

## DEDICATION

*As in the previous edition of my book, dedication must go to the many scientists and technicians responsible for producing new and improved hybrid plants.*

*We read so much about these new hybrid plants, even award them medals, but rarely do we read about the people who developed them.*

*The work of these dedicated people, laboring for many years before any progress is usually visible, has had a great effect on my Multi-Plant Method; higher yields and disease resistance are most important in gardens with limited space.*

*Ordering a packet of these great seeds is indeed so little for me to do. Given the treatment they deserve the returns are phenomenal, as you can observe in this book.*

*Needless to say, I am very grateful.*

# Acknowledgments

You won't find a list of "Courtesy of" photographs here; except for a single photograph sent in by a reader, all the photography in this book is the result of many color slides I've taken in the past years. However, there are two people who deserve recognition for assisting in publicizing my new method:

*Mr. Derek Fell, Director of the National Garden Bureau press service, who read about my method and was so convinced of it's greatness that he was determined to let other gardeners learn about it. He sent out a release proclaiming me "America's Champion Tomato Grower."

*Congressman John W. Wydler, who wrote about me in the *Congressional Record* July 25, 1975, explaining the many advantages of my method. He also added, "Mr. Klein has indeed come up with a fabulous method of growing tomatoes at a time when home gardening is increasing in popularity because of the rising cost of food."

Mr. Alvin Levin of William-Frederick Press, who helped me get my first edition of *Tomatoes: The Multi-Plant Method* in print.

My wife, Anne, who has lived through the difficult years of getting established, and who has also shared in the many exciting events that have taken place. Seeing my picture in many newspapers seemed to make up for the vacation time we didn't take, or the extra work she had to do entertaining visitors and answering phone calls.

My married daughter, Helene, who occasionally helped Mom with the typing when things got rough, and my son, Dean, who assisted outdoors when he was home from college.

Finally, I must include a dear elderly woman called Rose who lives next door. She, more than any other person, seemed to believe in what I was doing as she watched my experiments year after year. "Someday," she would repeat over and over, "people will read about you."

# Contents

# 1 Introduction

## 1 • Tomato Growing Is Not a Game of Chance

Many millions of home gardeners still live in hope that this year will give them better results than last and—hopefully—that big bumper crop. It is sad to see people working so hard year after year and getting such poor results. "Isn't the weather getting worse and worse? I didn't even come close to getting all the tomatoes that I did last year."

When someone says he had a bad year, it is really a cop-out for not knowing how. If commercial vegetable growers had to live in hope, we would have a serious food problem. The plain fact is that the people who supply us with our vegetables do not guess at it. Except for unusual weather conditions, growers know in advance what the expected crop will be—give or take a little.

The Multi-Plant Method represents a rare opportunity for the home gardener to grow tomatoes, and possibly many other vegetables, with the exact control heretofore possible only for professional gardeners. If you know the proper cultural procedure, and are supplied with a garden hose and various means of protecting newly set out plants, it is almost impossible to have a bad year—at least as far as tomato growing is concerned.

I can truthfully say that I have never had a bad year in more than thirty years of growing tomatoes. Now it is possible not only to get extremely high yields, but it is quite simple to duplicate these results year after year. As a matter of fact, each year I get higher production from the same amount of garden space.

# 2 • The Birth of the Multi-Plant Method

Over the past thirty years I have done research on growing tomatoes on a home level. (As a matter of fact, my experiments were with tomatoes only.) Having tried just about every known growing method, I am thoroughly convinced that the Multi-Plant Method is far superior to any other I have tried in the last thirty years.

After many years of using this technique, in 1974 I demonstrated in a book on tomato growing how four giant plants were able to grow out of a single square 20 by 20 inch hole in the ground or similar size box, and produce an estimated 100 pounds of tomatoes. Photos 1, 2, and 3 show some of my tomato units of the past few years. They will give you an idea of what is possible using the Multi-Plant Method. You may very well achieve similar results if you follow directions, but it is only fair to point out that not

*1• A ground unit of four plants, in late August*

*2• Another ground unit. Only about a quarter of this heavily-laden unit is visible in this photo*

*3• My best all-time yield. This picture shows only a section of one plant in a four-plant ground unit. The sixteen tomatoes shown weighed in at over ten pounds. Two near the 35mm film can (to give an idea of scale) weighed one pound each*

*all* my plants look as good as those shown in the photographs.

However, it is impossible to show the yields of all four plants in each unit. In some cases you will see one section of one plant bearing only a portion of its crop. If you multiply this yield by four plants, then consider that the plants continue to produce new tomatoes over the long four-month harvest period, you will get an idea how it is possible to get yields near one hundred pounds from one unit. I have been able to produce a lot of food for very little money, and it's all garden fresh and poison free.

Looking at Photo 4, that I refer to as "the good old days," it's hard to believe that I didn't develop this method many years before. Comparing the old to the new, the advantages should be quite obvious.

I used a sunny corner of my yard to grow twelve tomato plants. Preparing the ground was backbreaking work, since I dug down about 18 inches to get rotted humus material incorporated into the lower subsoil to encourage deep roots. Besides, the job had to be done in a single day or weekend.

***4• The "Good Old Days," back when I grew tomatoes the regular way***

*5• A box unit and ground unit—neater, easier to care for, and vastly more productive*

I now use that same corner to grow twelve plants using a few units, and I still have room for other vegetables. Small square ground units and my special boxes are now producing higher yields than are possible in much larger areas (Photo 5). Although the Multi-Plant Method is still confined to the home gardener, it is really showing signs of high efficiency even where labor is concerned. Once set up, these neat units require little work and watering and feeding are almost foolproof. There is very little to do except water, feed, take off a few suckers, and tie up the plants once in a while.

My first edition of *Tomatoes: The Multi-Plant Method* was published in 1974 by William-Frederick Press of New York. Because I had to pay for the book's publication, I was limited in the number of pages. This meant I had to condense my material more than I liked. Even so, the book got excellent reviews and I practically sold out the first edition. It described my new method for the first time, and my picture appeared in many newspapers after I was proclaimed "America's Champion Tomato Grower" by the National Garden Bureau. Later, the book was instrumental in getting Congressman John W. Wydler to enter my name and achievement into the *Congressional Record*.

It's hard to believe that all this came about because visitors to my garden were so impressed they demanded "How to" instructions. But actually, I think it all started when I was in high school. I had my own chemistry lab, darkroom, and workshop, with extra accessories including a medical microscope and a Leica camera, all paid for by a small baby portrait business and other odd jobs. As someone recently told me, "I think I can see the marriage of botany and mechanics in your work."

My interest in tomato growing started when I read an article called, "How to Grow Tomatoes Out of a Bottle of Water, by Chemistry." Having a real flair for chemistry and performing unusual experiments, I became quite interested. Indeed, the basic principles of hydroponics, or soil-less culture, have always had a great influence on my work.

I am not suggesting that you grow tomatoes in water or that my method is similar to hydroponics. Experimenters might wish to try their hand at it, but the average home gardener will find it too technical and expensive. But I think a little knowledge of this way of growing is very helpful and interesting. Once you understand the basic principles of this unusual way of growing vegetables, especially tomatoes, you will understand exactly how and why things grow.

I was quite surprised to learn that you could grow tomatoes without soil. You can grow, out of water, sand, or gravel, depending on your needs. As the article pointed out, if the plants can get air bubbled into the roots and you feed them the proper diet or chemical formula (all other conditions such as sun and temperature being correct), there should be no difference in the way the plants grow.

The chemical formula used is almost exactly the same as the solution fed to the roots of plants out in the field. By testing the solution from time to time, you can tell what chemicals need to be replenished. You can also keep the pH correct by adding acids or alkali solutions.

At the New Jersey Agricultural Experimental Station I saw a tomato plant growing out of a bucket of sand inside a greenhouse. A gallon of solution was so arranged that a single drop fell continuously on the plant every few seconds. Or the plant was hand fed every few hours. The plant had reached the top of the greenhouse, was still growing, and had already yielded many pounds of tomatoes. In reading about this experiment, I found that sand provided the roots with a good holding medium. Also, there was never too much or too little solution surrounding the roots.

The roots were supplied with large amounts of oxygen as water gravitated downward, pushing tiny bubbles of air in front of it. It is probably from sand culture experiments that I got the idea of supplying large amounts of oxygen to the roots. Confining them, giving frequent waterings (plus feedings of course) seemed to give me that good growth I saw many years ago at the New Jersey Agriculture Station. However, I went one step ahead and placed four plants where one normally would grow; using a larger soil bed of course. I found this increased efficiency; also less space was used, and less labor was required to maintain the plants.

But even more important, instead of the special solutions that were added a few times a day, I used one watering a day and one feeding a week of an inexpensive fertilizer.

It wasn't until high school that I made up my mind to

pursue a career in science and hopefully in medicine. However, Uncle Sam had other ideas, and after I returned from Europe in 1945, I became an animation cameraman working on cartoons, and later T.V. commercials. But I still had an urge to experiment with chemicals. I can recall how my young marriage almost abruptly ended when I spent more than a week's salary to install a water culture experiment to grow tomatoes in my small back yard, shortly after the war. But it wasn't very long when I realized that this highly technical way of growing vegetables was not for the home gardener, and returned to mother earth.

Photo 6 of my wife on a ladder, about 25 years later, shows my first Multi-Plant box experiment. Actually, it was a surprise to my wife. For weeks I had been working on

*6• My wife tending the first Multi-Plant box unit*

something in the basement. Anne didn't know what it was and I was almost afraid to tell her. Finally, when the unit started to take shape, I told her I was making something to grow tomatoes in. "Isn't it kind of big for a tomato plant?" she asked. I told her it was an experiment to grow four plants together, for people with apartment balconies or roof tops. "I'm going to put wheels under it so it can be moved from place to place—possibly even to follow the sun."

A few weeks later I asked my wife to lend a hand to move the box (approximately 2 feet by 2 feet by 2 feet) out into the yard. One of the neighbors was a little curious as I rolled the unit near the fence adjoining our property. For many weeks the box could not be seen from the next house. Friends who visited me asked what it was. But as the months passed, vines loaded with tomatoes started to reach for the sky. Curiosity reached its peak when I sent my wife up on a ladder to pose for the picture shortly after Labor Day. Tall tomato stakes were not something new in my yard, but these were extra high because they were attached to the box. As neighbors looked over the fence they were amazed to find four monstrous plants all growing out of a box! It was another idea of mine using mechanics and it was a success the first time out. A few years later, a visitor looking at this box turned to his wife and said, "Look—a tomato machine."

In my many years of experimenting with tomatoes, I have used many unusual devices that helped make things easier or produce higher yields—or make a neater garden. My Multi-Plant Method now incorporates a long list of new things never before seen in gardens or books on vegetable growing. However, even before I started growing four plants together I had many unusual gadgets and ideas at my disposal. Many of these ideas are probably my own, since I hadn't seen or read about them at the time.

Working in a small corner of my yard—all my wife allowed me in those days—simplicity was my most important goal. I tried to eliminate things I usually added to the soil; surprisingly, it proved possible. I never grew the exact same garden each year; at least one change had to be made. And usually, each year brought higher yields from my same small plot.

To make a neater garden or produce higher yields, I tried various mechanical ideas. This led to the use of specially constructed stakes, Masonite box sides, and of course to various new boxes I designed.

Years ago before leaving for work, especially on real hot days, I first had to water my plants and check for caterpillars. Walking between wet sticky plants with your good clothes on is a job for a dedicated gardener only! Now I service about fifteen units of different vegetables before I leave for work. Each unit gets about one minute of the hose. I circle around my tomatoes quickly to check for caterpillars. While I am growing many, many more pounds of vegetables, I don't get messed up walking between rows of plants.

By now, you can understand why I gave up the old way. "But why should I grow four plants together when I have plenty of room?" someone asked.

How long will it be before space becomes a premium and good water becomes scarce? In some parts of the world space is already a problem. And in most regions of our country, good drinking water is already in short supply. Doesn't it make sense to conduct experiments *now* on how to solve these shortages?

"But look at all the space you have between your units. Why are you wasting it?" someone asked me. It is quite obvious that if I can grow one hundred pounds of tomatoes in a space of about five feet by five feet, I could grow many more units together if the need for it arose.

But when people space individual tomato plants too far apart, I feel this is truly poor economy where space is concerned.

A young girl escorted her grandfather into my garden to show him the great results I get. Despite his reluctance, she pulled him in and stopped near a ground unit of four plants that had almost one hundred red and green tomatoes hanging, looking like a Christmas tree. He paused and said, "You're in big trouble. Your plants are too close together." Indeed, crowding many plants together *is* bad, unless you have a purpose as with my method. Actually, less space is not the *only* reason for Multi-Plant growing, nor even the most important reason. A bed of deep, rich soil is responsible for the rapid growth. It is not likely that you would dig and prepare your soil as deeply as I do, thus you are not likely to have as good results. (I wouldn't take so much time and effort to grow a *single* plant.) So it all works out quite well. As I told someone, "It's all a matter of placing the right things in the right place at the right time."

The greatest advance appears to be with my box unit, which of course has received the most attention. High yeilds are now possible on an apartment balcony or any above-ground site. In a recent article I wrote, "My new compact vegetable gardens will soon replace the present way of growing vegetables, at least as far as the home gardener is concerned."

Today I own a small motion picture company, and I must admit that I now lead a double life. No longer just a hobby, tomatoes have become even more important to me than motion pictures. And my work is no longer limited to tomato growing.

Now that Prentice-Hall is publishing this revised edition, there is enough space to "tell it the way it is," plus room for much new material. In the few years since I wrote my original book I have improved my Multi-Plant Method: constructed new and better boxes, enjoyed super yields from early variety tomatoes, gained a better understanding of soil preparation, and many other things. But it is foolish to think that we have reached our goal in growing tomatoes or other vegetables. Each year a slight change has either produced greater yields or eliminated problems. And I'm certain there are still greater things ahead.

Since you will be spending many months, from early spring to late fall, watering, feeding, caring for, and worrying about your tomato plants, doesn't it make sense to learn the only practical way—my Multi-Plant Method? Grow one or two of my ground units, and at the same time grow a few plants the old way. In the following pages, I will describe in detail, and with clear photos, how simple it is to get good results.

My prediction is that even if you are an experienced gardener accustomed to producing high yields, the Multi-Plant Method will give you red tomatoes many weeks before your regular crop. Secondly, you will learn that growing tomatoes can be done in a neat and simple way. And if you are an inexperienced gardener, you will be amazed at the results.

## Organic Gardeners Invited

Although my present methods do not strictly adhere to organic standards, there is no reason why organic gardeners can't adapt them. If you're having good results at present, you should do even better. Because of the many requests by organic gardeners, I am planning an all-organic unit for next year, and should have all the details available soon.

*Myself with one of my Box Units*

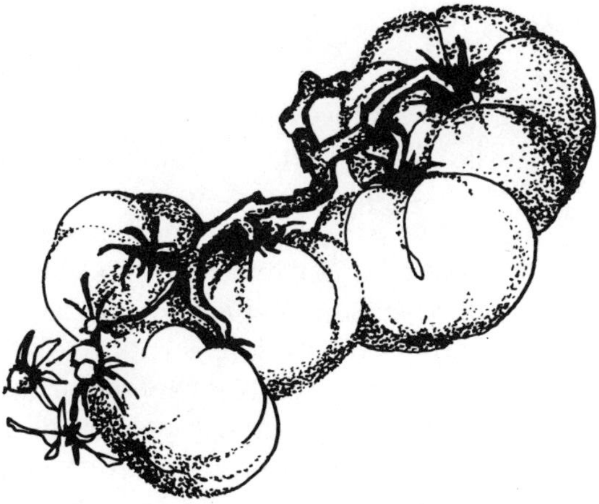

# 2 Getting Started

## 3 • Preparing Organic Material for Your Soil Bed

The very first thing you should do is to take time to read this book which will give you a very good idea of what to expect. It will be to your advantage to read it well in advance of "tomato fever time," between May 1 and May 15, when tomato growers rush down to their local garden center and grab anything they can find. This panicky atmosphere will not produce the best results, since you are sure to cut corners in your attempt to get your tomato plants into the ground as soon as possible. Whereas a well-planned unit with plenty of time to prepare it can't fail—at least my way.

The material in this book represents more than thirty years of my research. In order to enjoy the benefits of this great technique, you must follow directions as closely as possible. Once you start, it's all or nothing. Things are not critical, but there is no room for substitutions. You can't, for example, forget to add mulch, feed the plants when you feel like it, water the units all day long, or completely forget to water except when you think of it.

I'm certain that you have heard the word "compost." Compost is any combination of organic material that is allowed to decay. It is then incorporated into the garden to become a very important part of good soil.

Compost should not be confused with manure that contains nutrients, or plant food that can sustain growth. Usually other materials containing phosphorus are used with manure to form a more complete fertilizer. Manure, unfortunately is quite expensive in my area. Luckily I found a way to make use of nature's products without using expensive manure.

Compost that has not been aged does not have many nutrients in it to sustain good plant growth. It is used mostly to loosen the soil and act as small sponge to hold water and plant food. (As a substitute you could use peat moss.) By season's end, almost all the organic material in the Soil Bed will be reduced to ordinary humus, adding a certain amount of richness to it.

Although compost making is not new, there has been a greater interest in its use over the last few years. There are entire books on the subject, since *aged* compost can be used to supply all the nutrients needed to grow things. After all, that's how nature fertilizes trees and other wild growth.

Compost is important for the Multi-Plant Method, and there are now many different formulas used to prepare it. Some so-called experts advise various layers of leaves, grass clippings, etc., to be built up or piled on top of each other. There are also chemicals for sale that are supposed to speed up the decaying process so the final humus will be created much sooner.

But I have my own simple way to prepare compost.

While some gardeners fill their compost piles with just about everything that can be called organic, such as leftovers from the kitchen table, I am more selective. I have two sources of compost. My main and probably best supply comes from weeds dug out or pulled with soil still in contact with the roots. I deposit them in a hole shaped almost like a bathtub, dug in the far corner of my lot. Actually, the longer the trench, the better, since it will hold more material.

No grass clippings are put on this pile; I prefer a heavier type of material. Besides weeds, I also use pieces of sod and edgings from the lawn. It is not necessary to add soil to the pile during the summer, and definitely do not add fertilizer or lime. This may be contrary to what you have heard, but I prefer to add my own fertilizer when actually preparing the Soil Bed.

In the fall, when you feel there will be no more to add to the pile, you may cover it with an inch or so of the best soil available.

My other source of compost or organic material comes from leaves. People waste much time and money bagging up leaves when all you have to do is bury them. I use a large area where flowers once grew, about 5 feet wide and 15 feet long. I dig out the center about 1 foot deep and 3 feet wide, add leaves, walk on them to press them down, then add some soil. I repeat the process until I have about three layers of leaves, then cover the whole pile with soil.

I also dig out the soil from my ground units and fill these with leaves and soil. Another excellent source of material is the scrapings from the lawn in the fall. These must be covered with soil for the winter.

Grass clippings can be used if nothing else is available. Not having tried them, I would imagine that you could deposit all your clippings on a pile after you mow your lawn each week. Actually, the bottom of this pile is best, since it would have been decaying for many months.

*7• The compost heap dug up in early spring. It can be left in a pile like this, ready for use, or dumped back in the hole*

## Digging Up the Pile

As early as possible in the spring, start digging up the various buried materials. I work as early as late March since I have a sandy type of soil. You must separate the various types of materials. In the case of leaves, especially, this is very important. The leaf material should be turned over and placed on a pile. It is wise to turn over the pile a few times to really mix the soil and leaves well.

The unit with weeds has to be turned only once. Dig it out into a pile and them dump it back into the hole again. Or it can be left as a pile. Now you are ready to prepare the Soil Beds some time in April.

# 4 • Picking a Site for Your Units

About the most important consideration in deciding where to grow your plants is of course sun. To get really good results you will need a minimum of five hours of sun. If you have less, try it anyway; you might be surprised.

What if you have tall trees in your back yard? As I see it, here are your alternatives:

1. Use your front lawn.
2. Build a box and find a sunny spot for it.
3. Build a box with wheels under it, and move it a few times a day to follow the sun.
4. Try to trim some of your trees to let sun in.
5. Chop your trees down.
6. If all fails, move.

There's another important point to consider in picking a site. If you have a choice, do not grow too close to bushes whose long roots might invade your Soil Bed. But you may have no choice. If you plan to grow on the south side of a tall tree, where you would get no sun on the north side, you will have a serious problem with heavy roots in the ground. Ideally you should use an above-ground box in sunny spots. However, in many of my photos you will notice a box-like effect around my ground units. These are pieces of Masonite raised above the ground as a kind of basin to retain water until it seeps into the soil.

It's only my theory, but I believe that by using deep sides I restrict or contain the roots. Water and fertilizer are similarly contained so that I will always have a deep, heavy root system so vital in growing deep-rooted plants like tomatoes.

In Diagram 1, you see that my four plants take the same root space as one single plant would normally. As water gravitates downward, it is poor economy to have only one set of roots when you can service four sets with the same water.

You have a choice of not using any Masonite. You'll still get good results. The depression may wash away in time as you apply water, but do not be afraid to try it. You can use a modified version—just deep enough to go down a few

*Diagram 1*•

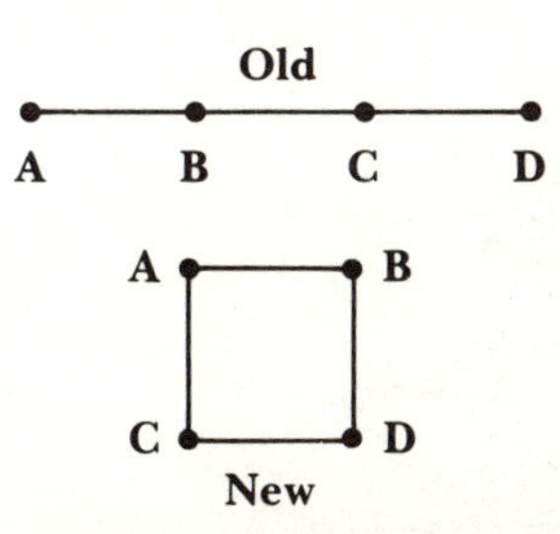

inches below the ground—but still maintaining that box-like effect. Or you can use pieces of Masonite that extend to the bottom of the hole on all sides. If the hole is close to bushes and trees where there is a great concentration of roots in the ground, the deep sides are a must. If you wish to restrain outside roots, you must also dig out the soil around the Masonite after each season, and allow the hole to stay open all winter to discourage roots. If you left things alone, the second year you would have large roots working into your Soil Bed.

## Masonite Squares

If you plan to use either the full size or shorter size, you should order 1/8 inch Masonite from your lumberyard. Tempered-both-sides is best, but many places carry only tempered-one-side. For the full, deep sides you should order four pieces 20 by 20 inches. If you plan to use the modified version, order four pieces 20 by 10 to 12 inches deep.

It is not necessary to protect the tempered side of the Masonite, but it is quite important to add a wood preservative to any dull or unfinished surface. This will allow you to use these pieces for a number of years before they decay. Wood preservative can be bought at lumberyards, paint, and hardware stores. You will also need this liquid for making your tomato stakes, described later.

*8• A ground unit neatened up with concrete blocks*

I also added concrete blocks around my units so it would look almost like a flower garden (Photo 8).

One word of warning: In rare cases, you may have the type of ground conditions where drainage is so bad that an ordinary rain can sit there for days. If this is the case, you should consider means of alleviating this situation—possibly by digging an extra deep hole and lay in a foot or so of sand or gravel before adding the good soil.

Of course the real solution is making a box.

# 5• Box Units

It is true that my tomato box is receiving most attention, but if you have a regular garden a box is not necessary. This method works well in a ground unit or box, and I don't want you to go to the expense and bother of making a box. It is important to know that the box is more critical when it comes to water. During the hot summer months, and with a heavy load of tomatoes on four giant plants, you can't skip a single day's watering without the plants wilting.

However, if you feel that you would like to impress your friends and neighbors, especially if you are handy with tools, don't let me discourage you. Although I have plenty of good ground space I now have four boxes, and I make a new one each year to grow other vegetables. Even in the hot days of summer, one good morning watering will last until the next day.

By late summer, the plants in the box are higher, so reaching up top might be a problem for some.

Nevertheless, I must admit that I am starting to favor the box over the ground units. It makes such a fast start early in the summer because the sun is allowed to warm the soil and the roots while the ground is still cold and clammy.

It is protected against heavy rains and floods that cause damage to ground plants. Being high it is easier to work with, especially for older people who can't bend over comfortably. The box works well with many other vegetables and I sometimes wonder if one day I will have an all-box garden? Can't you just see your wife serving out on the terrace and going from box to box collecting fresh vegetables? Such a garden would be the ultimate in growing vegetables, especially in affluent homes. It would of course be automatically watered the same as your lawn so that you could leave it for as much as a week, but at least for long weekends!

## Newspaper Box Versus Deluxe Unit

It's been many years since my first Multi-Plant experiment in a box, and I have since changed the design many times. My research is split between a colorful unit intended to match the décor of the most affluent and a box that is easy and inexpensive to make.

*9• The Newspaper Box, taken about July 15*

*10• The Deluxe Unit –without newspaper insulation, and without plants to show detail*

One is called the Newspaper Box (Photo 9), and the other I refer to as my Deluxe Unit (Photo 10). Looking at the two photographs you will understand why I sometimes like to call the Deluxe Unit the Permanent Box. You must be realistic about it; the solid outer plywood box is quite durable compared to the 1/4 inch Masonite of the other. But don't get the impression that the Newspaper Box is flimsy; it really isn't. My Newspaper Box has been in use for three years and given good care will probably last many more.

To help you decide which box to make, ask yourself the following questions: Is it important to have a nice looking unit, or are you only interested in growing tomatoes? Are you handy with tools? Is the cost of materials a problem? The Deluxe Unit is a joy to behold and will always be welcome in the proper surroundings. But from now on all my new units will be Newspaper Boxes because they are cheaper and easier to make.

## Wheels Versus No Wheels

Both boxes can be made movable by adding a set of dolly wheels under them. On the colorful Deluxe Unit, especially, I feel that wheels should be standard equipment, almost like power windows on an expensive car. But don't let me talk you into something you don't need!

# 6 • The Newspaper Box

If you follow my instructions, this box can be made to last many, many years. If you try to cut corners it may not last one season. Ideally, the box should be made many months in advance; work on it a little each night. Except for the time it takes to apply wood preservative and paint, the box can be put together in a few hours once the parts are cut to size and some holes are drilled.

This unit is mainly a square Masonite box approximately 20 inches by 20 inches by 20 inches. I use a 19 inch box, 20 inches high. It uses less soil than a 20 inch and the results are very good. You can obtain all the pieces necessary for construction pre-cut to size from your lumberyard. You will have only to put the box together and paint it.

## Masonite and Wood Cut to Order

One-quarter inch Masonite, tempered on one side, is used to construct the box. If you can obtain Masonite tempered on both sides, it is even better. However, it is often not available, more expensive, and really not essential.

The dull or untempered sides of Masonite should be protected with a wood preservative and painted. It will then be almost as durable as a piece of steel plate.

Order four pieces of 19 by 20 inch 1/4 inch Masonite. If you are a perfectionist like me, and you have your own electric saw, cut two pieces 19 by 20 inches and two pieces

*11• Odd pieces of wood are screwed together to form the base. The two cross pieces are needed beneath the box for drainage*

18-1/2 by 20 inches. This will make a perfectly square box, but it is not necessary.

You will also need four pieces of 2 by 3 inch wood. These are used to reinforce the sides to which the stakes will be attached. The bottom is made of 3/4 inch plywood, although I have also used odd pieces of 3/4 inch lumber cut to size to make 24 by 24 inches overall. If possible, order exterior plywood, although I have used the interior type successfully since I add wood preservative anyway. Plywood that is finished on one side only is all that's necessary and it's cheaper.

The box can be set on dolly wheels or on strips of wood to allow for drainage. I will explain about the dolly wheels later. The strips can be pieces of plywood or regular lumber 3/4 inch thick by 24 inches long and 3 to 4 inches wide (Photo 11). If you are using odd pieces for the bottom, 4- to 5-inch pieces might be better because they are what holds it all in place. If you do use odd pieces, place them *tightly* together so that soil will not fall through.

You will also need corner pieces to hold the box together. Order four pieces of 1-1/8 inch square lumber 20 inches long. (This lumber will be more like 1 inch square.)

*12• End view of Masonite box from the Deluxe Unit which is similar to the Newspaper Box*

Finally, you will need strips of wood to hold the Masonite box to the base (Photo 12). Ideally, the same 1-1/8 inch square lumber is best, but I have also used 3/4 inch square. Order eight pieces 7 inches long. While you're at the lumberyard you can order the wood for the stakes that you'll need come June if you wish. You'll find the specifications in Chapter 27.

## Miscellaneous Screws and Nuts

To fasten the box together, you will need at least fifty flat-head wood screws, #6 by 1 inch. You might as well order a box of one hundred. You will also need two to three dozen flat-head wood screws, #8 by 2 inches to hold the Masonite box to the base, and a dozen or so #8 by 1-1/2 inch, flat-head screws to hold the pieces under the base. In addition, buy eight 1/4-20 round-head machine screws 3-1/2 inches long. Get twenty nuts to fit, and a dozen steel washers about 1 inch in diameter. Extra nuts are needed to make stakes slant outward.

Dolly wheels can be bought at a good hardware store.

*13• Plywood base drilled for drainage*

They should be 2-1/2 to 3 inches across and must swivel. To install them you will need 8 round-head machine screws with nuts 1/4-20 by 1-1/4 inches. Two for each wheel is enough.

## Necessary Tools

An electric drill with a few bits is practically a must. The bit sizes I use are 1/4 inch for the holes for the screws to hold the stakes, and 1/8 inch to make pilot holes for the wood screws that hold the box together. A countersink attachment will allow you to make the flat-head wood screws flush with the surface. A good screwdriver will also help but it is almost obsolete. I have an attachment to my drill that allows me to screw the box together in minutes instead of hours.

## Predrilled Holes

Holes should be drilled before you add the wood preservative. Twenty-five to thirty 1/4 inch holes must be drilled in the 24 by 24 inch plywood base (Photo 13). I use a countersink on both sides to clean out the hole. If you're using odd pieces for the base, add preservative first. Put it all together, drill holes, then pour the liquid into the holes.

If you're using dolly wheels you should also predrill the holes for them. Lay them out on the base on the four corners and drill 1/4 inch holes. Actually, the holes can be 5/16 inch. The point is, there should be clearance.

## Predrill Holes in Side Pieces

Two 1/4 inch holes must be drilled for the screws that hold each stake. It is important that these holes be *exactly* the same distance apart on both the boxes and the stakes, so the latter will be interchangeable. To ensure this, you should make a template, guide, or pattern (see Diagram 2).

*Diagram 2*

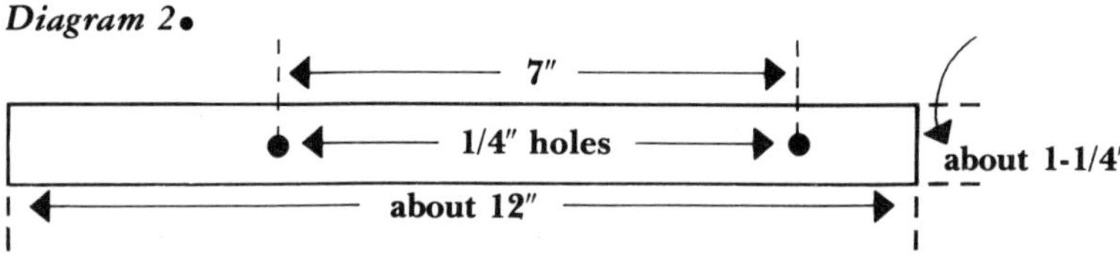


**Template or guide for drilling holes in 2″ by 3″ pieces**

I made mine from a piece of steel about 1 inch wide and about 12 inches long. If you can't find a piece of steel, you can make the template or guide from Masonite or similar material. Drill two 1/4 inch holes about 7 inches apart. Clamp this to your pieces of wood and use the two predrilled holes in the template as a guide for your bit.

Note that in Photo 15, the 2 by 3 pieces do not reach the top of the box. This shouldn't be! They *should* reach the top. Measure down about 1-1/2 inches, lay on the guide or template and drill two 1/4 inch holes. Once the two holes are drilled you must make a dummy or clearance hole using a wood drill 1/2 inch to 5/8 inch—halfway into the

*Diagram 3•*

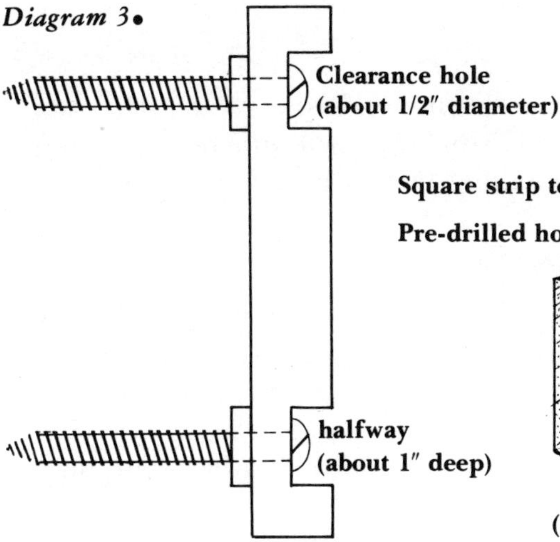


*Diagram 4•*

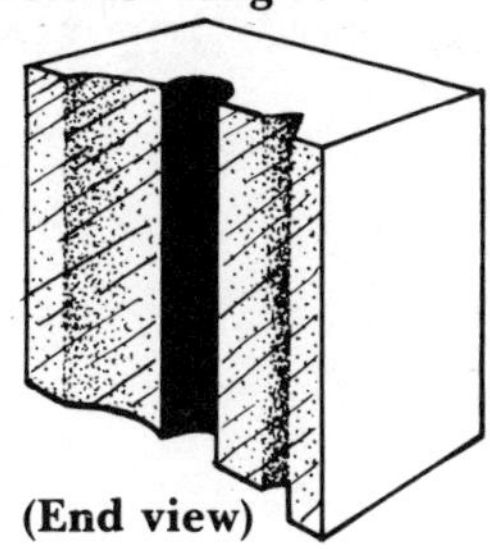


piece of 2 by 3 wood (see Diagram 3). This allows you to use shorter and less expensive screws and makes a flush fit. After you have added wood preservative, you tighten the screw in place before attaching it to the side.

Finally, the square strips holding the box to the bottom should have holes for the screws predrilled in at an angle (see Diagram 4). You will need a bit slightly larger than the #8 by 2 inch screw. It is best to countersink the holes.

## Adding Wood Preservative

Once all the holes are drilled, the parts must be protected with a coat of a wood preservative. This provides added protection under the later coat of paint. Wood preservative is bought at lumberyards, paint, and hardware stores. It's a clear liquid easily applied with a 2- to 3-inch brush. Use a little caution and work outdoors. Use gloves if possible.

The large pieces of Masonite need be coated only on the dull sides—the edges and back. It is not advisable to coat the shiny side, since it will face inward where the roots will come in contact with it. All the wooden pieces must receive a coat of this liquid, including the inside of the holes. Allow it to dry, then assemble the top part of the box. Wash your hands well after handling this impregnated wood.

## Assembling the Box

After making a few of these boxes I have found that a certain order of operations works best. For example, two corner pieces should be added to one of the sides first. Use the 18-1/2 inch side if you have it. Place the two square pieces on a table and lay one of the Masonite sides on top

*Diagram 5*•

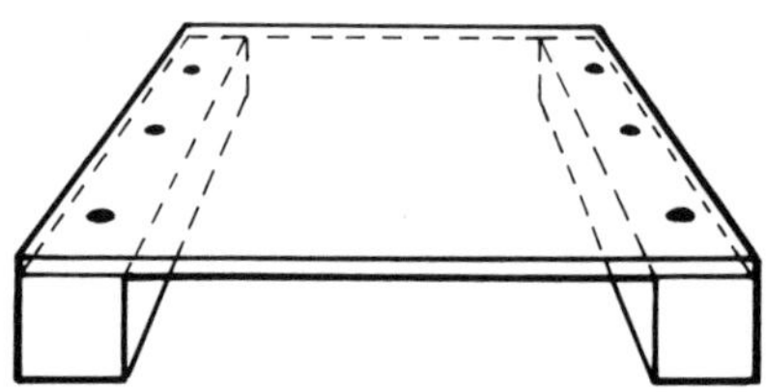

**Screw first two corner pieces to one side**

of it, shiny side down (see Diagram 5). Carefully line up the Masonite flush with the square pieces and add about five wood screws. Using the 1/8 inch bit, drill in about 1/2 inch to make a pilot hole. Countersink the hole so the flat-head screw will be flush with the surface. After my first one, I drill all the holes, then using my drill attachment, I tighten all the #6 by 1 inch screws in mass production style. Repeat on the other side using the other piece of Masonite. You now have the four corner pieces in place.

To add the strips along the bottom, flip the piece of Masonite and lay the strips under. You should lay a similar piece under the opposite end, so the bottom will lie evenly and parallel with the table. Allow for the 2 by 3 piece before you screw these in place. Use the same procedure as above, securing the strips with #6 by 1 inch screws.

*Diagram 6•*

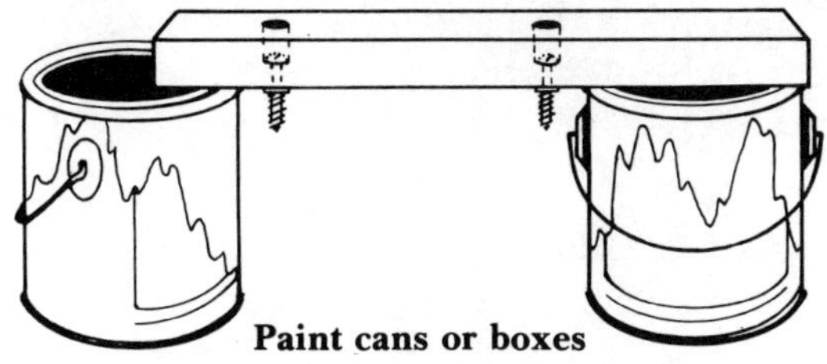


**An aid to screwing 2″ by 3″s onto Masonite sides**

Finally, add the 2 by 3 pieces to the sides. This is a little tricky, since you must first put the long machine screws in place on the 2 by 3 pieces. Put the screw in the hole and secure tightly with a nut. A washer is not necessary. When the two screws are in place on all four pieces, these will be added to the sides. Before attempting to attach these pieces, raise them up on two paint cans or boxes, so the screws will be facing down; you can now apply pressure when securing with #6 by 1 inch screws (see Diagram 6). If possible, clamp the two pieces together on the end before screwing together.

*Diagram 7•*

**Screwing sides together (top view)**

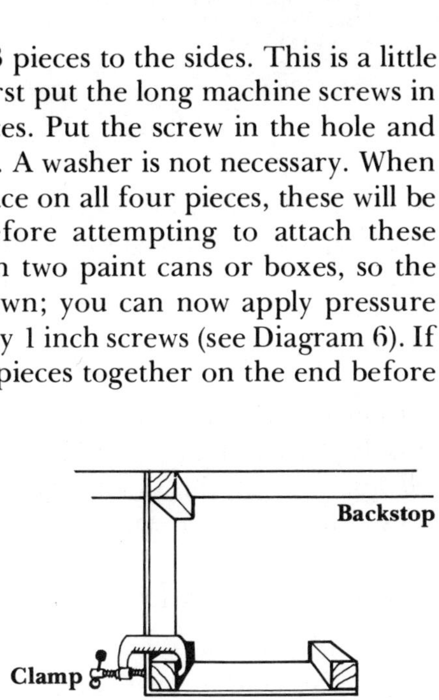


Now it is just a matter of screwing all four sides together to form the box. It is best to work on a table by a wall that you can push against. If you don't have such a setup, you can clamp a strip of wood on your work table to act as a stop (see Diagram 7).

Stand up two of the sides to form an L. Unless you have a clamp you should get someone to hold the two pieces

*14• Box with 2 by 3 pieces added*

together, at least while you put in the first screw. Now continue as before, using about five screws on each side. Then take the other two sides and repeat the procedure. Finally, form the box and add the final screws. Before securing the final side, make sure the box sits evenly on all sides.

I must emphasize that this box must *not* be waterproof: its only purpose is to contain the soil. And while it must not fall apart when the plants are loaded with tomatoes, it doesn't have to be a work of art (Photo 14).

## Ready for Painting

The paint to use is a high-gloss enamel designated for interior or exterior use. It is available at lumberyards, paint, hardware, and department stores. I usually get a quart of dark green. Give all surfaces a coat of this paint and allow to dry. If you bought the wood for stakes you can also paint them now or wait a few weeks. The enamel need only be dry to the touch before making the final assembly to the base. If you are using dolly wheels, now is the time to put them on. Use 1/4 by 20 round-head screws 1-1/4 inches long. Two on each wheel are enough.

*15. Affixed to base, painted, and ready for planting*

## Final Assembly

Place the base on the table or floor, then take the box and center it on the base (Photo 15). Using your 1/8 inch bit, drill into the holes a little. Then secure all sides with #8 by 2 inch screws. If you used 3/4 inch square pieces, you can probably get by with 1-1/2 inch screws. As an added protection I add a few #8 by 2 inch screws to the base of the 2 by 3 pieces. Using your 1/8 inch bit, drill in at an angle about 3/4 inch up and make some pilot holes. Do not drill too far into the base, just enough to get you started. Use a larger bit near the top so that the screw will clear. Then add the screws.

## Bringing It Outdoors

Before putting soil into the box, leave it out in the rain a few weeks to allow the paint odors to evaporate. If it doesn't rain you can help by hosing the inside of the box every few days. Some people will tell you that you shouldn't grow anything in containers that have been painted. But I have had many years of excellent results.

As hot weather approaches, wrap about a 1-inch thickness of newspapers around the box and protect them with clear polyethylene. The newspapers insulate the box and—more importantly—protect the roots of the plant from the hot sun (see Chapter 34).

# 7 • The Deluxe Unit (A Box Within a Box)

The inside of this Deluxe Box is almost exactly the same as the Newspaper Box, except the 2 by 3 side pieces are not needed. The stakes will now be attached to the outer plywood box. Compared to the Newspaper Box, measurements will have to be more exact. Otherwise you might end up with twisted boxes that will not fit well.

## Pieces to Order From the Lumberyard

The inner box will be made of 1/4 inch Masonite and will be exactly 19 inches square, 20 inches high. For it you need:

1/4 inch Masonite tempered on one side:
2 pieces 19 by 20 inches
2 pieces 18-1/2 by 20 inches
8 square corner pieces 1-1/8 by 20 inches (four for inner and four for outer)
4 square pieces to hold the Masonite box to the base each 1-1/8 by 17 inch square
2 pieces of 3/4 inch plywood 24 by 20 inches
2 pieces of 3/4 inch plywood 22-1/2 by 20 inches to form the outer box
The base is 3/4 inch plywood 24 by 24 inches

Thus, you will be making an outer box of 3/4 inch plywood 24 by 24 by 20 inches. If you wish you may add 2-1/2 to 3 inch dolly wheels that swivel. If no dolly wheels are used, use two strips of any wood about 3 inches wide and 24 inches long under the box for drainage.

## Screws, Etc.

As for the Newspaper Box, you will need a box of #6 by 1 inch flat-head wood screws and about four dozen #8 by 1-1/2 inch flat-head wood screws. (You might as well get a box.) About four to five dozen #8 by 2 inch flat-head wood screws. (Might also get a box.) Finally, eight 1/4-20 round-head screws 3-1/2 inches long with 20 nuts to fit and a dozen washers.

## Predrilled Holes

For the inner box, the square pieces that hold the Masonite box to the base must have holes predrilled at an angle (see Diagram 4). About twenty-five to thirty 1/4 inch holes must be drilled in the base. Also holes to take the dolly wheels if used. Use the countersink on both sides to clean out the hole.

Using the template described for the Newspaper Box, measure down 1-1/2 inches from the top of the plywood side and drill two 1/4 inch holes (use a clamp to hold the template). Make sure the hole has enough clearance to insert the screw through later, after the wood is painted.

The plywood box will be put together with #8 by 1-1/2 inch flat-head wood screws. It is wise to make these holes before adding the wood preservative. Using a bit slightly bigger than the 1-1/2 inch screw, drill holes along the sides, finished side up. Measure in about 1/2 inch and drill five

*16• Inner box for Deluxe Unit*

holes on each side. You will be actually screwing into the square corner pieces later. Countersink the holes. The base will also be screwed on, but it is better to do that later without using predrilled holes. Sandpaper the rough edges of the 3/4 inch plywood.

If you are preparing the stakes now, predrill the two 1/4 inch holes using the template, measuring in about 1-1/2 inches from the end. As with the Newspaper Box, give all surfaces a coating of wood preservative except the shiny side of the Masonite.

## Assembling the Inner Box

Refer to section on Newspaper Box. Lay two square corner pieces on the table. Take the 18-1/2 inch piece of Masonite and lay it on top, shiny side down. Line it up evenly and secure with screws. Flip over the Masonite and add the square strip along the base as before. Repeat with other 18-1/2 inch Masonite side. Taking the 19 inch side piece, add it to the 18-1/2 inch piece forming an L and secure with screws. Repeat with other two pieces.

Finally screw the box together (Photo 16).

17• *Outer plywood box*

## Outer Box

This box is put together the same way as the inner box, but it is much heavier and harder to work with. Again, lay two square corner pieces on the table. Take the 22-1/2 inch piece of plywood, finished side up, and lay it on the two pieces. Line it up evenly. Before adding the #8 by 1-1/2 inch wood screws, drill into the square corners a little way with the 1/8 inch bit. Repeat, using the other 22-1/2 inch piece of plywood. Then, as before, take one of the 24 inch pieces and from an L; secure with screws. Finally, screw the entire outer box together (Photo 17).

## Ready for Painting

As with the Newspaper Box, all surfaces must be given a coat of paint. However, if you want to get fancy and give your box a two-tone effect it will require extra work.

I used yellow and light green for the two outside pieces, and dark green for all other surfaces. You will need two coats of paint for the two outside colors. For best results, be sure the edge of the base matches the side.

## Final Assembly

When the box is dry, secure the two 1/4-20 by 3-1/2 inch machine screws on the sides of the outer box with a nut. You will not be able to do this later, so make sure they are screwed tight. If you plan to add dolly wheels, do so now, or make sure you added the pieces under the base to keep it raised for drainage.

Then place the base down and secure the inner box to it, using the predrilled holes and #8 by 2 inch wood screws. It must be centered. You must now turn everything upside down. Using assistance if needed, turn the plywood box upside down. Then take the base and inner box and flip it over on top of the outer box and center it. You are now ready to screw the base on. I use #8 by 2 inch flat-head screws for this. First, make clearance holes through the base, using a larger bit. Then drill into the other piece a little using the 1/8 inch bit, then countersink. Finally, add all the screws. Caution: You will be screwing into 3/4 inch lumber and it must be accurate.

## Newspaper Insulation

The way the boxes are now, the outer box will act as a barrier to prevent the hot rays of the sun from striking the inner box except when the sun is overhead. While this

*18. With insulation sealed with Masonite strips*

might be protection enough, there is no point in taking a chance and it is best to do it the right way. By the time the hot days arrive you should have newspaper insulation between the inner and outer box. But you can't just leave it that way, or water will soak the papers and make a mess. I use strips of 1/8 inch Masonite large enough to make a neat covering and screwed into place (Photo 18). If you have a better idea, use it. This insulation can be left in place from year to year.

# 8 • Growing on Balconies and Rooftops

Some of you will want to set up a box on a balcony or rooftop in the heart of a big city. Your problems will be somewhat different from using a box in the suburbs where gardening accessories and soil are readily available.

A young married couple who had read my first book sent me a color slide of their tomato box growing on an apartment balcony (Photo 19). These people never grew a vegetable in their life, and on their first try produced a monster!

Those using a box on a balcony may purchase peat moss and topsoil if regular materials are not available. A number of my readers have told me that a 50-50 mixture works well, although I have never used it. Both are available in garden supply stores in most big cities. They usually come in small size bags so they are easy to carry, though you might have to pay extra for this convenience. Very roughly, you will need about 10-11 10-quart pails of soil material. When it is thoroughly soaked, it will pack down. You will also need at least two 10-quart plastic pails with handles. And a few gallon bottles, preferably plastic, with screw caps.

*19• The Multi-Plant Method on an apartment balcony*

## Best Location

Ideally, a southern exposure would be the best location for balcony units, although anything between east and west would probably give good results. Your worst exposure is north, although even such a balcony might get some sun, depending on the way the building is laid out.

It is extremely important to have a portable unit so that you can rotate the box, especially if there is a balcony overhead, thus exposing all the plants to the sun evenly. If you are entertaining guests at night, you can push the box out of the way in a corner. And in case of a strong windstorm, you can pull the unit close to the wall for protection.

# 9 • Preparing the Soil Bed

When you have picked a good sunny spot for your plants and have all the necessary materials on hand, you are ready to prepare your unit. If using a box, it should be ready. First of all you should get a bag of 5-10-5 fertilizer (Photo 20). The 50-pound bag, though heavy, is the cheapest. Five pounds per unit is required during a season. While shopping you should also get a small bag of ground limestone. Very little is used, and only for the initial soil preparation (Photo 21).

*20• 5-10-5 fertilizer*

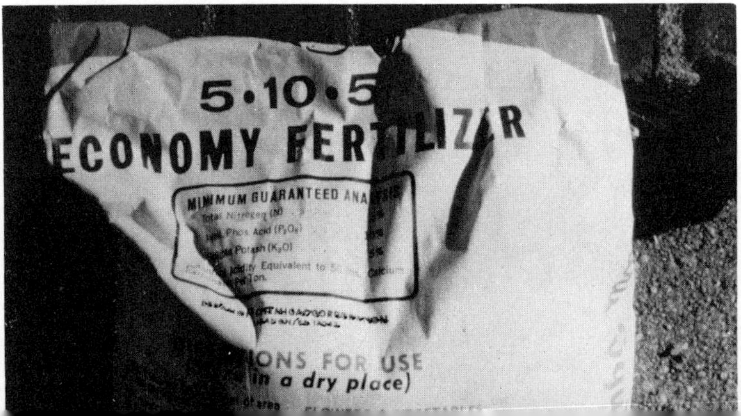

*21• Pulverized or ground limestone*

## No Soil Test Needed

In my many years of growing tomatoes I have found that soil tests were not needed. Fall and winter rain and snow leach most nutrients out of the soil; what little is left is not a problem when computing your fertilizer applications in the spring. (In fact, you should *try* to wash out the fertilizer still left in the soil.) It must be understood that the Multi-Plant Method means working with small compact units as against the conventional way where you dig up an entire area. As such, you have better control of your soil. A pail of this and a pail of that, with a little 5-10-5 and lime seems to work very well without first making a test.

It would be wise to have your ground unit holes dug out in the fall, then all you have to do is fill them. However, you can dig out the hole a few weeks, or even a few days before planting time. Normal time is mid-April. The hole should be 20 by 20 inches and about 18 inches deep. If you are using the Masonite pieces they can serve as a guide to measure the exact size needed (Photo 22).

*22• Hole for basic ground unit*

I mix my units about three weeks ahead of planting, allowing plenty of leisure time, but the units are ready to be used once the muddy effect is gone, possibly in a day or two.

First, you will need the materials to make your soil mix. In Chapter 3 I explained how to prepare compost or organic material so important to your Soil Bed. However, if you can't yet scrape up enough decayed organic material from your yard, try your town's dumping grounds. They usually have piles of leaves which you can have for the taking. Of course the stuff under the pile is what you want most, but almost anything will be usable.

Your second choice is peat moss, available at garden centers. You will also need topsoil.

There are so many variations of soil mix that it is not easy to cover all the different possibilities. Of course if you are already working with good garden soil, all you need is compost added to it, plus 5-10-5 and lime.

If you never grew anything in the spot you choose, chances are that it contains stones and hard clay, or it may be all sand. In either case, only a small part or possibly none of it, should be used.

If you live in the western part of the country, you might check with your local garden center or your neighbors to find out if the soil is alkaline. This is fairly rare except in certain parts of the country, but it is a very serious situation. These soils contain so much alkali that it is almost impossible to grow vegetables in it without making corrections. In such cases a soil test might be justified.

If you are so unfortunate as to have such soil, I suggest that you try to build up a supply of new soil. Since you are using small, compact units, you can purchase topsoil and peat moss and continue to use them according to directions in the book. This might be far easier than to try to correct your present soil.

When digging out soil, the top layer is usually pretty good and should be put to one side. The subsoil is usually not usable and should be removed to a spot in your garden where you can just spread it out. If you have replacement material, by all means use all new soil. If you don't, you can use some of what you have.

If you are growing near the beach where there is nothing but pure sand, it is important to keep your good soil isolated. Possibly your sand contains salt, which can cause problems. Here again, the deep Masonite sides (with no bottom) for ground units are a must. A portion of sand can't really hurt, as long as it's kept to a minimum.

As far as clay soil goes, I have been getting good reports from users who added quite a bit of organic material to it. Heavy clay soils, when mixed with plenty of compost or peat moss can make a pretty good mix that does not dry out too fast. Although I have not used such material, there is no reason why such a mixture will not give good results. However, you may have to judge for yourself what a good mixture is and possibly add some sand. In some cases the addition of some builder's sand (not beach sand), if readily available, should be used to lighten the soil so that water will drain better. But I must emphasize that the mix is not critical; almost any combination will work.

## Mixing the Ingredients

Your soil preparation is not like that of a commercial farmer planting acres of crops. Instead, it's blending a deep loose mixture with the same measuring care of a housewife baking a cake. You cannot mix the soil properly directly in the box or ground unit. When I prepare my Soil Bed I use a large area where I can lay out and mix the different ingredients. If you are working on a lawn you can put down a large tarpaulin or something similar. Or you can work in a wheelbarrow. For balconies it is best to

*23. Topsoil plus compost on tarpaulin, ready for mixing*

lay a piece of large plastic on a concrete floor and thoroughly mix small portions of soil at a time.

Start by dumping a pail of whatever organic material you have available. Then add a pail of your present soil or topsoil (Photo 23). The general rule of thumb that I use is one pail of compost or other organic material to one pail of regular soil or topsoil. This is for most normal situations. However, the compost you make has some soil in it, while peat moss does not. In such a case, I think you should use less than half the amount of peat moss and slightly more soil.

Even after many years of incorporating organic material into it, my soil is still on the sandy side, so I use a little more than half compost. In the last year or so, I have changed to a heavier type of soil, adding more organic material or compost and keeping the sandy part down to a minimum. The box, in particular, should have a little more heavy organic material to hold water longer, while the ground unit can take more of your regular soil if you're short on the organic material.

I usually dump four to six pails before I start to mix. You can make as much or as little as you want at one time. Mix it up well with a rake, then lay it out flat, a few inches thick, as in Photo 24. Then lightly sprinkle 5-10-5 fertilizer

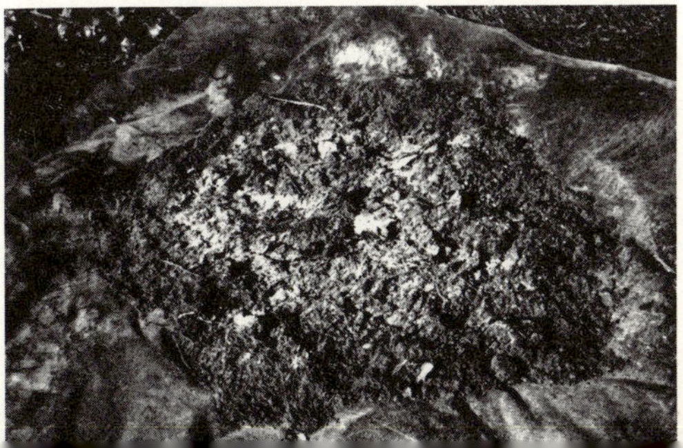

*24. With lime and fertilizer added, before mixing with rake*

*25• Ground unit with Masonite sides and Soil Bed filled in*

and lime over the soil; the addition of compost will almost always add acid to the soil, so it is standard practice to add lime. But the small amount of lime used is so inexpensive that it should not even figure into the consideration of cost.

Mix it all well and again spread it out as before. Repeat the application of 5-10-5 and lime, mix well, and it's ready to be dumped into the hole or box. (Two separate applications of fertilizer and lime are recommended because it is much safer than trying to mix a large amount all at once.)

Warning! It is extremely important never to allow a blob or large amount of fertilizer—mixed or unmixed—to remain in one spot. It will absolutely cause trouble later when the roots of a plant strike it. For this reason it is not advisable to use anything but fertilizer that is dry and can easily be spread. Old bags containing large pieces are risky unless they are broken down into a powder first. The use of lime is not as serious. But here again, don't drop a handful in one spot without raking it thoroughly.

If you are using a ground unit with deep Masonite sides, they should now be put in place making a neat box (Photo 25). The pieces should protrude above ground a few inches, but they do not have to make a perfect fit. Do not cut notches in the Masonite. (Square concrete blocks are optional.)

If you are using the shorter pieces of Masonite, wait until you have filled the hole a little more than halfway before inserting them to form a neat box. Dump the mixture into the hole or box, and then prepare more until your soil almost reaches the top.

It might be wise to mix up a little at a time until you fill the unit. After you have thoroughly soaked the soil bed it will settle down to about four inches below the top of the box or Masonite. This is just perfect. It will allow for a few inches of mulch and a little extra space for water to collect.

When the soil is prepared above ground and dumped into the hole or box, I find that I get a better average of good producing plants. In some cases one seedling will do better than another because of different genetic factors. But discounting the heredity factor in the seed, all plants should grow equally well in a soil where all the nutrients are fairly equally divided, as against the hit-or-miss technique used to prepare large areas. With my method it is rare to find an outstanding plant here and a weakling there.

Once you have had good results you should have no problem the following year. You know the type of soil you are working with, and it is just a matter of following the same directions given in the book.

# 10 • A Soil Bed for Organic Gardeners

An organic gardener who read my first book wrote that he got good results using my method. He mixed some manure with his compost and added more on the Soil Bed surface after the plants set tomatoes.

Before attempting to start an organic Soil Bed or grow seedlings, you should refer to these sections and make the substitutions where necessary. It should be quite simple to make certain substitutions when preparing the soil. When good manure became difficult to find, I got annoyed and substituted 5-10-5. Nevertheless, you can still use manure; as any experienced gardener knows, a fairly good crop of tomatoes can be obtained using manure and good soil. However, as books on organic gardening often mention, manure does not contain enough phosphorus and so is usually supplemented with additions of this chemical element. Bone meal and rock phosphate are two materials used to increase the available phosphorus content so valuable in hastening the ripening process.

Not having used manures for many years, I am not claiming to be an authority. I used to obtain it by the bushel or in what might be called unprocessed form, but it is available at garden centers, dried or processed, in bags. The rough stuff obtained from farms usually contains straw, sawdust, and similar matter and wouldn't be as potent as the dried product, thus allowances should be made when using it. By the way, you should never use fresh manure until it is well aged and takes on a dark brown color.

It should be understood that for best results, besides the manure, some form of decayed material or compost must also be added to the soil. In other words, manure is not a complete substitute for the organic material I recommended in the section on soil preparation. Conversely, compost alone does not supply enough nutrients to sustain good growth. It *is* possible to make a compost that can supply the needed food requirements, but this requires a special technique of preparation and usually takes many years.

You can prepare the compost following my instructions, or you can use your own method. I only ask that you do not add any lime or fertilizer to the pile; I prefer to add these materials when I prepare the Soil Bed. If you are using unprocessed manure, a little extra can't cause any problems, but you should dismiss the idea that you can grow tomatoes in a concentrated Soil Bed of the stuff. As mentioned elsewhere in the book, you need only enough plant food to reach the next stage—setting the tomatoes. If too much fertilizer is added at this time, no matter what type, you will sacrifice the buds. The rest of the plant will develop faster than the tomato producing blossom or flower, resulting in a later harvest.

Even more serious problems can result from a heavy concentration of manure near the roots. Many years ago, I recall, I placed a shovelful, unmixed, about a foot below where I set out my tomatoes. Things were fine until the plants got taller and the roots started to reach down. Then suddenly the growing tips of the plants twisted up like corkscrews. It didn't take me long to realize that if a little is good, a lot is *not* always better.

# 3 Growing Your Own Seedlings

## *11 • Seeds*

Whether you buy plants or grow your own you should know which strains are best for your needs.

I have read claims that the old favorites tasted better than today's varieties; that Jersey tomatoes, or those of some other state, taste the best. I have grown tomatoes continuously for thirty years, and have actually tried the old and the new. In my opinion, today's varieties will stand up to any old-timer. When I pick a spotless, vine-ripened three-quarter pounder from my vines, early in July, there is no tomato in the world that is superior.

Raising your own tomato plants from seed is not only economical, but you also have a choice of the best available seed in the country. Moreover, your plants are available exactly when you need them.

*26. Tomato seeds, greatly enlarged*

Nowadays, tomato seeds are cataloged almost like parts of a machine. Photo 26 shows a few seeds from many different packets; as you can see they all look alike. Yet one will produce cherry-sized tomatoes, another one-pounders, a third orange-colored tomatoes, while a fourth will emerge as an Italian-style plum tomato. However, there will be many that look almost alike, except that the plants that grew them will not be the same. One will have the ability to produce more than the other. Another will be resistant to disease, while yet another will produce almost crack-free tomatoes.

The next important thing to consider is the word "hybrid," appearing more and more in catalogs. Hybrid indicates that a particular seed, whether tomato or other vegetable, was produced by crossing two different plants. This will be explained in more detail in the chapter on pollination. These newly developed plants not only have the ability to produce more, but in many cases, new improvements are bred into them.

Hybrid plants seem to have more vigor and can produce more than the old favorites, which were of the self-pollinating type. Hybrid seed costs more, since extra labor is involved to produce the seed, but the extra cost is surely well worth it. Today, all my tomato seeds are hybrids only.

As far as I am concerned, it makes no sense to save seeds from hybrid tomatoes or other plants—you are a sure loser if you do. In the first place, the seed from a hybrid plant no longer contains the original F-1 qualities. But even if you tried to use the seeds from tomatoes, as old-timers do, you are really fooling yourself. The small amount you saved on buying seeds could never begin to equal what you would lose in production by not using hybrid seed each year.

Tomato seeds and plants generally fall into three main types:

1. *Early* The early variety tomatoes have a very definite place in the home garden, and I have devoted a special chapter (39) to them. These tomatoes are smaller in size, but are ready a few weeks before the large midseason type. You will want to learn about my "super yields from early varieties" that can produce 250 or more tomatoes, many about the size of a golf ball, starting in mid-June and continuing for more than a full month, supplying your family with all the tomatoes they can eat until the big red ones are ready.

2. *Midseason or Main crop* Though any tomato plant will grow well Multi-Plant style, the midseason types seem best for our units. They stake well and produce many large tomatoes weighing three-quarters of a pound or more. All the large tomatoes mentioned in this book are midseason.

Since my midseason plants are hybrids, I felt I might as

well get an extra bonus and also get disease resistance. Today almost all the newly developed tomato plants are resistant to Verticillium and fusarium wilt and are so denoted by the abbreviation VF. I must admit that I still use some early varieties that are not VF resistant. I like their qualities so much that I make exceptions in a few cases.

Not having tried all available seeds, I can still comment on a few midseason varieties I have had good results with. All are VF resistant hybrids: 'Supersonic,' 'Burpee VF,' and 'Better Boy VFN.' The 'N' indicates resistance to nematodes also.

Nematodes are tiny worms found in the soil; they can be a problem where they exist. There are now seeds which are claimed to be resistant to this problem (discussed later in Chapter 38).

Clearly, three varieties are not a complete listing. I have only a limited amount of space for my experiments, and I can't test all available seeds. If a seed company feels that they have a good plant for my units I hope they make it known. I would be glad to try other seeds that are claimed to be outstanding.

Finally, be on the lookout for All-American Winners. When such a designation is indicated for a tomato or other vegetable, it is a sure sign that you have a good seed. Surprisingly, none of the three above seeds are A.A.W. For some reason there haven't been many awarded in the midseason type.

3. *Late* Actually the word "late" is confusing, since a midseason plant will also produce late tomatoes. A late tomato is simply larger in size than most midseason ones, thus it will ripen later. An example is a Beefsteak, rarely used by today's experienced gardeners. It was very popular many years ago, producing a large size tomato. However, many tomato growers still look for the sign Beefsteak. As one dealer told me, "I don't put out the Beefsteak

sign to fool my customers. Most want a Beefsteak type of tomato, the large ones, and that's what I'm giving them."

I suggest that you grow more than one variety. You can see for yourself the different features of each, and benefit from the advantage of having a better supply of tomatoes available, since different varieties ripen at different times.

## 12 • Keeping a Diary

It is important to know what you did in the garden so that you can easily duplicate your results or make corrections the following year. Some things are very important to record so that you won't have to guess:

"I used uh—uh—what was the name of those tomatoes? I wish I could remember what the man said, they were pretty good." "Was it the end of April or early May that I set out my plants?" "I had my first red tomato very early, sometime in June, I think. But it couldn't be, we used them for our first barbecue early in August." Important information should be recorded, not only for your own needs, but to prevent making a fool of yourself when you talk to an experienced gardener.

You don't need a book. All you really need is a sheet of paper and something to write with. You can use two methods—record weekly events, or as important things happen.

Here are some important things to put down:

- The name of the seeds and where you purchased them
- How you mixed the soil for your seedlings and when you started them indoors
- Where you placed them, temperature, and when they broke through or sprouted

- When you transplanted to individual containers and when they went out for hardening
- How you prepared the Soil Bed
- When you set them permanently into the Soil Bed
- The approximate day and night temperature
- How long you covered them
- When you removed the plastic greenhouse
- Your first bud noticed
- Your first flower
- When you started and stopped using the liquid to set early tomatoes
- When you broke off your first sucker
- First tomato noticed
- First tiny insects noticed
- When you started to feed
- When you put in your permanent stakes
- Your first tomato turning red
- Your first tiny green caterpillar
- When you were late adding your mulch, but will make sure to add it earlier next year
- When the weather started to get warm—your first heat wave above 90 degrees
- How long it lasted
- General comments about unusual weather or how your plants are growing
- When you added extensions to your stakes
- First frost
- Comments about end of season

# 13 • Starting Seedlings Indoors

Milk containers slit in half to make trays to start seedlings is probably another of my ideas. (See schedule for starting seeds in Chapter 19 and 21.)

*27• Milk carton tray for seedlings and needlenose applicator for precise watering*

As you can see in Photo 27, I use a tray made by slitting a half-gallon container the long way. If you cut it exactly in half using a razor blade, one piece will be waterproof. For two trays, get another half-gallon container, or use the upper part of the first one by taping the end up. No holes are needed for drainage. In the absence of a milk container, almost any similar tray may be used. I suggest using two trays.

For very small seeds, a sandy soil usually makes it easier for the sprouts to break through. My soil is a mixture of sand that I usually dig out from my subsoil here on Long Island, to which a little topsoil is added. I take in a pail of it in late fall and allow it to dry out thoroughly near my furnace. Even if you don't have such a mixture almost any soil will work, but it is best to try it ahead of time.

Place about an inch or more of this dry sandy material in the tray and level it off. Pick out all the larger pieces of twigs and stones. I use a small block of wood, just large enough to fit inside, to level and tamp down this dry material. No fertilizer is used in the flats to germinate the seeds.

## Laying Out Seeds

Most hybrid seed comes in a packet containing thirty to fifty seeds. These you should distribute evenly in both flats. The greater the space between seeds the better. You

*28. Seeds correctly spaced and matchsticks or toothpicks for estimating proper depth*

must use a different container for each different variety. If you are using only one flat do not put them closer than 1/2 inch. Since I grow many plants, also many varieties, I am well supplied with enough seed and I usually give many of my spare plants to neighbors.

Once the seeds are in place, I sprinkle about 1/4 inch of the same dry soil over them. To make sure I use exactly 1/4 inch, I set toothpicks in the soil, allowing 1/4 inch to protrude (see Photo 28). When the toothpicks are covered, I know I have the right depth.

Using the wood block, I again tamp down and level the soil. I use an old catsup dispenser or any similar needle nose applicator for watering instead of other crude ways. Carefully and thoroughly wet the soil, making sure that you have an estimated penetration of about 1/2 inch. It isn't serious if you have more, but at no time allow the flats to get waterlogged.

## Place in Warm Spot

Contrary to popular belief, it is not necessary to place seeds in a dark place. If this were so, no seeds would germinate outdoors. The idea of placing the seeds in the dark probably came from the fact that a warm furnace room is an ideal place for seeds to germinate. I usually put the flats about three feet away from the furnace. This is not at all

*29• Wait until a dozen or so sprouts have emerged before moving the flat into the sun*

critical; you can set your trays in a warm kitchen too. A sunny windowsill is fine when the sun is shining, but not so good when the temperature drops on those cold March nights. I have found a fairly steady temperature of from 70 to 75 degrees F. will send sprouts up in about five days. The containers must be wet a little every day, and even twice a day if the surface dries out. All you need is enough water to wet the surface.

## Move to a Sunny Window

Once the sprouts have broken through, they must be moved to a sunny window. However, I sacrifice a few and wait until a dozen or so have emerged before bringing them up (see Photo 29).

A window facing south is best, although east and west exposures will produce usable plants. North, with no sun, is almost useless.

To give me more room for my plants I don't use simply a windowsill but an extension—a piece of 1/4 inch Masonite, measuring about 18 inches wide, and the width of the window, with two stands to hold it up. One of these stands, which is almost toppleproof, can by itself hold a waterproof box with eighteen plants in individual containers. Also see Chapter 15 on how to make waterproof boxes as well as the stand.

*30• Sunlight immediately strengthens the tiny seedlings*

While I have always used a sunny windowsill to grow my plants, if you have been getting good results using the special fluorescent lights very popular in the last few years, decide for yourself. These tubes, available in almost every electrical store that sells lamps, come with directions for distance, time, and so forth. I do feel that if you don't have a good sunny window, you should consider them.

The first few days are critical. When watering, you must be very careful not to knock down the tiny seedlings.

In a few days they will become stronger and will be out of danger (Photo 31). But for now, water the flats daily.

As soon as the seedlings get their second or true leaves (Photo 32), they should be transplanted to their individual containers.

# 14 • Transplanting to Individual Containers

About one week before you are ready to transplant your seedlings, you should have your containers filled with prepared soil. After many years of raising seedlings, I have chosen the familiar empty quart milk container for my

*31• First leaves*

*32• First true leaves*

*33• Quart milk container for individual seedlings*

plants. I do not feel justified using the larger half-gallon container. The little extra that it buys is not worth the extra space, soil, and weight.

I save my quart milk containers all year to have them ready when I need them. Friends can possibly supply you with more. I use the bottom 4 inches and cut off the four corners with a razor (Photo 33), leaving four triangular holes for drainage instead of making holes in the bottom. This idea has proved very effective.

Unlike most pots, these containers hold more soil and can produce a large root system. Besides, they do not dry out fast and are simple to cut away when the plants are ready to set into the Soil Bed. A large root system is immediately in contact with the damp rich soil, which contributes to the almost instant growth so necessary to produce early red tomatoes.

## Styrofoam Drinking Cups Can Be Used

I am not suggesting that you forget about the milk containers and use these drinking cups, but I have used them with pretty good results even though they do not hold nearly as much soil as the milk containers and they can easily tip over, damaging the plants. But they are very simple to break away when plants are ready to place into the Soil Bed.

When using these containers make four holes with a sharp pointed pencil, not in the bottom, but at the base of the side, almost like the milk containers.

## Preparing the Soil Mix

Almost the same mix is used as in the outdoor units except that smaller pieces of organic material are used. As I said, my soil is somewhat sandy which encourages a heavy root system. However, heavier soils are used in most of the potted plants or flats sold at garden centers. A mixture of topsoil or your regular soil plus some organic material added to it will do fine.

If you purchase your materials, some topsoil with a little peat moss works well. If a little sand is added, so much the better. But don't go out of your way and *don't* use beach sand. A little 5-10-5 and ground limestone will also be needed, as with all soil mixes. The lime tends to neutralize the acid from the organic material. As I mentioned earlier, I take in pails of the various soils and allow them to dry out all winter. You can't go looking for soil in March. But this does not mean only dry soil can be used. Simply allow the soil to dry out a little if it is very wet.

## Lay It Out on Newspapers

Working in my basement, I lay down a few thicknesses of large size newspapers on my worktable. I take a few pots of the regular soil, then a few pots of the organic material and mix it up, spreading it out about 1 inch thick. I sprinkle a little 5-10-5 over it, followed by a little lime; mix well and spread it out again. I then repeat with the 5-10-5 and lime so that two applications of fertilizer and lime were used. This is a lot safer than trying to add the full amount at once. Finally, I mix it again, fill up my pots or containers, give them a thorough soaking, and set them aside to be used later.

For organic gardening, the same substitutions could be made when preparing the soil to be used in the containers to grow seedlings. In place of the 5-10-5 you should add the manure and bone meal or other form of phosphorus. Two to three times the amount of manure should be used to compensate for the higher rated 5-10-5. A little extra bone meal can't really hurt, since it is so slow acting. Add lime as instructed.

*34. Transplanting with a teaspoon*

## Ready to Transplant

By now the little seedlings have their second or true leaves and are ready to be transplanted. Actually if the plants get a little taller they will be easier and safer to transplant. In fact they can be moved any time, even after they are quite large. If transplanted when only a few inches tall they are easier to dig out; as time passes the flat becomes a jungle.

Put your prepared container of soil on newspapers. Then, before you begin transplanting, it is important to water the flat. Not too dry or too soggy is correct. The soil should cling to the roots when seedlings are dug out. Using a teaspoon, dig out about half of the soil in the container (study Photo 34). Then look for nice healthy plants and using your spoon as a trowel dig well under the plant to get both soil and roots.

Soon you will get to know how much soil to dig out of the container, according to the size of the seedling being moved. Slowly move it to set inside the half-filled container.

Take the loose soil and place it carefully around the plant, leaving about 1 inch of space at the top. It is wise not to cover too much stem, or the plant will take longer to recover. You then gently tap the soil in place with your fingers and set aside (Photo 35).

*35• Properly transplanted seedling . . .*

*36• . . . with mulch added*

If you are not happy with a certain move, especially if you damage the roots so the soil falls off, discard the plant and try again. If two plants come out together do not separate them. Pack soil around both. Then with two fingers and a sharp fingernail, break one off at the base. When all have been moved, water just a little with your needle-nosed container and keep them out of the sun for a few days. They can stay there in the basement.

Before you take them up into the sun you will need to add about 1/2 inch of wet mulch to the top; this prevents the soil from caking or becoming hard (Photo 36). Don't omit this step or your seedlings will not grow well. Your mulch can be peat moss, but I use dried sod roots from my compost pile. Place the material in a pail and soak it well. Then place a little carefully around each plant.

The first day up from the basement, leave the plants in a room where there is light, but not in the sun. The next day or so, place them in full sun. This will be about three days after transplanting. Water the plants every day, about an ounce at first, increasing to about two ounces when plants are almost ready to go outside.

Since plants bend to the sun, they should be revolved each day.

If there are any seedlings still left in the flats after the rest have been transplanted, do not discard them. Add soil to fill in the holes in the flats. Keep them watered and in the sun—not in your sunniest spot, but at least a little. These plants can then be used in case a plant does not grow well or dies of damping-off disease. You can also give some plants to your friends.

A cooler room will produce buds faster and therefore, earlier red tomatoes. I tell you this only if you have a choice, but do not be too concerned, since the difference is not too great.

If you see a few plants suddenly keel over, it is probably due to damping-off, a fungus disease. My suggestion is not to worry about it unless you lose many of your seedlings. In all my years I have never lost more than one or two, and therefore have not done anything about it. One recommended prevention is to place the soil before planting in a hot oven for a time to kill all the fungus spores. I guess that allowing my soil to thoroughly dry out all winter also kills them, since most such fungus diseases thrive on damp conditions.

Now is the time to look into making a neat waterproof window box to hold containers. The waterproof windowsill box will hold eight to ten plants. However, you can set the plants on the windowsill without the box. Place some

*37 and 38• The waterproof Window Box, in two different sizes*

newspapers under them to sop up the little water that leaks out.

## 15 • Waterproof Window Boxes and Toppleproof Stand

There are actually two different types of boxes that can be made, depending on the number of seedlings needed. For only eight or nine plants you can make one that can be placed on the windowsill. However, if you need as many as eighteen seedlings, you can construct another kind of box that has served me well for many years.

I used 3/4 inch plywood on the larger box only because I had the odd pieces; 1/2 inch plywood would be fine. The windowsill box can be made of even lighter material, possibly a 1/4 inch base with pieces of 1/8 inch Masonite for the sides.

The actual size of the larger box is made according to certain limits. The windowsill box should be made in accordance with the dimensions of your window.

Eighteen quart milk containers measuring 2-3/4 inches square will take up exactly 16-1/2 by 8-1/4 inches, but a little extra room will help. If it is to be placed in your plastic greenhouse, it must be just short of 20 inches if using the ground unit, or less than 18-1/2 inches if using the box.

If you plan to get it inside the plastic greenhouse, the critical dimension is the long side. But there is no point making the width less than eleven inches.

I am not going to give you instructions on how to make a box. But when it is completed I suggest you make it waterproof and possibly paint it to match the surroundings, as my wife made me do. To waterproof the box I used a caulking material that comes in a roll. I merely squeezed it into the corners and painted over it.

As the plants start to branch out, shortly before moving them outdoors, I usually take three plants out of the larger box to allow more room, and place the three near the windowsill.

## Toppleproof Stand

Many years ago I designed a Toppleproof I stand to hold the heavy box with eighteen plants. I have since made many and use them for holding a small television set, a potted plant, and two more supporting a piece of Masonite used to hold many seedlings near my sunny window.

The stand is made from pieces of 2 by 3 inch wood plus two square pieces of plywood. First you must measure the distance from floor to windowsill. I have found it's better to make the stand a little higher than the sill rather than lower.

The top and base are identical and can be 1/2 inch plywood 12 inches square. Besides the center piece, you will also need four other pieces 2 by 3 by 6 inches long. The length of the center piece is the distance from floor to sill, minus the thickness of the top and base.

***Diagram** 7A•*

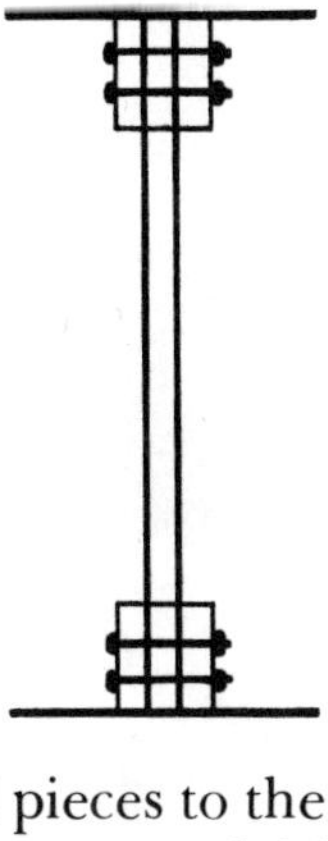

Screw the two small pieces to the top and bottom, allowing for the center piece; use fairly long flat-head wood screws. Then insert the center piece and drill long holes for long 1/4 inch machine screws to hold it all together. To make a neater job, I made clearance holes so both ends of the screw can be recessed. I then filled the holes with caulking compound. As shown in Diagram 7A, two screws are needed above and below. Large clamps holding the three pieces together will help, but are not really necessary. I painted mine white.

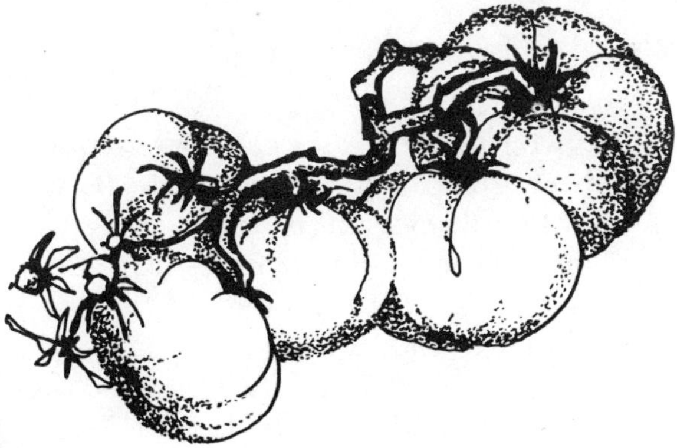

# 4 Moving Your Plants Outdoors

## 16 • Hints About Getting Early Tomatoes

People who visit my garden in late June are surprised to see large green tomatoes, and in recent years red tomatoes from the early varieties. They usually ask, "Is there a trick to getting early tomatoes?"

The answer is yes, except that I like to think of it, not as a trick, but good cultural or gardening practice. According to my diary on gardening, which goes back many years, June has not been great to set tomatoes. But if you already have tomatoes on your plants June is a good month to fatten them up. For many years, the early part of June has had long periods of rain followed late in the month by heat waves above 90 degrees. High temperatures and long periods without sun are both bad for setting tomatoes. But once July arrives the rains seem to stop, and early July has been great weather for tomato plants.

This assessment is based on the average condition over many years and is not necessarily the type of weather you'll actually get. But commercial farmers do use these figures to plan ahead. All you have to do is make sure your plants that are already in the ground have small tomatoes on them by mid-May, or at least early June. If they don't, there's a great likelihood that your first small tomato will not appear until early July.

You have probably read that you shouldn't set out tomato plants until the weather is warm and settled. Not necessarily. To have little tomatoes on your plants by mid-May, you must adhere to my schedule very closely and plan to have well-established plants ready in time.

Contrary to what you may have read, a tomato plant has the ability to withstand cool weather without damage; note that I did not say freezing weather. This characteristic lets you try various means of getting the plant established in the Soil Bed well in advance of the normal planting time. And for this reason you can expect to get a large red tomato on your table by about July 4, or weeks earlier with the smaller early varieties.

# 17 • Hardening Plants

When a tender plant that has been growing in a warm windless atmosphere is suddenly placed in the cool windy outdoors, it's almost like you going outdoors in winter without clothes. This chapter explains the proper way to make this transition without damage to the plant.

In order to harden a plant and make it ready to be set out, you must remove much of the water it contains.

Strong winds and hot sun will both remove the water, but much too fast. The process must be gradual or the plant will wilt. Hardening your plants actually accounts for earlier tomatoes since this procedure forces the buds. Once in a cooler environment the foliage will not advance as fast, thus giving the buds (and eventually the flowers or blossoms) a chance to develop faster.

To get an early start, I have designed a simple plastic greenhouse costing pennies. Using it you can safely get your plants off to an early start without any concern. Protection from the weather was heretofore only available to the most affluent gardener using expensive greenhouses. Anyway, while a greenhouse is not really practical during the summer months, this unit is. It's practical for my method only because of the square format of my Soil Bed or box. Protecting the unit with simple, inexpensive plastic walls that can be covered at night makes it possible for you to get an extra early start not possible using conventional or ordinary gardening methods.

# 18 • The Plastic Greenhouse

To make the plastic greenhouse, you need four 30 to 36 inch long sticks—one for each corner. I used 3/4 by 3/4 inch lumber (see Photo 39). They are sunk in the ground with about 18 inches remaining above. Note that the sticks are placed *outside* the Masonite. Try to place them so they

*39• Plastic Greenhouse for ground unit: first the corner stakes . . .*

40• *. . . with plastic and central stake added*

will not be more than 20 inches apart from each other. This will permit you to place a piece of 1/8 inch Masonite, 24 by 24 inches, on top of the four sticks. You can add wood preservative to the dull side though it will only be used for a few weeks.

You are now ready to add a piece of plastic to the outside of the sticks, forming four walls. I use clear polyethylene four mils thick, sold in rolls at garden centers. Cut pieces 18 inches wide and long enough to go around all four sticks.

Or you can use shorter leftover pieces and go from stick to stick (Photo 40).

The plastic is now fastened in place with staples. Use at least five or six staples on each stick since you will be raising the plastic up and down. Actually, this shelter is a temporary, makeshift affair, but it works. You will be ripping the top sections of plastic from the staples during the day and putting them back at night, but you don't have to staple it back each night.

You will also need another stick, the same as the corner sticks, for the center to prevent the Masonite from sinking

41• *With Masonite lid and stones*

42• *Tarpaulin for night use*

in when a heavy stone or two is placed on it to keep it from flying away in a strong wind (Photo 41). Finally, to use the unit to full advantage, it must be covered at night for a few weeks. You will need a tarpaulin or similar covering for this. I use 6 by 8 foot plastic ground covers sold for a few dollars at most camping supply houses (Photo 42). You'll also need to locate a few large stones to hold the plastic in place.

Most of these things can be used again and again so the initial outlay can be spread over many years. Absolutely nothing is critical about making the greenhouse or using it. It is not intended to be completely sealed from the weather.

## 19•Schedule for Starting Seeds with Plastic Greenhouse

Start seeds indoors March 5 or March 1 in a 68 degree temperature where the plants will grow. Note: you still start the seeds in a temperature of 70 to 75 degrees.

You can buy your plants between April 24 and May 1. If possible, you should select those that are almost a foot tall,

*43• Greenhouse in use on a box unit. Inside are plants still in their containers*

but without tomatoes on them. Then place them in the plastic greenhouse.

Place your house plants (in containers or pots) outdoors in the plastic greenhouse for hardening, from April 24 to May 1. At this point they should be about 10 inches tall. If the plants are much taller or shorter, you should make a correction next year. You either started too late or too early. Quite obviously, if you can cover your plants to protect them from freezing temperatures or damaging cold winds you can set them outdoors earlier. So, if you plan to set them out earlier, you have to start them or buy them earlier.

Between April 24 and May 1 in my area, I put my plants in my plastic greenhouse. You do not guess at this move or even worry about weather conditions outside up to a point.

It is best to bring the plants out at night and completely cover them. A cloudy or rainy day is also a good time to set them out. Plants are placed in the greenhouse still in their containers and possibly the waterproof box. Or you may sink the containers a few inches into the soil, but don't bury them (Photo 43). They can be covered completely for a few days, but if the sun is shining allow air to enter the greenouse. Study Photo 44 and you will see how the plastic

*44• Adjusting the plastic on a ground unit to let air in during the day*

is pulled down 4 to 5 inches while still covering the rest of the greenhouse. (Please note that even though it appears that the sun is shining into the opening, you should make the opening on the north side so that no sun will shine in. The picture was taken on the sunny side only because it would make a clearer photo.)

After two or three days you can remove the covering during the day. However, if your plants wilt, cover them again, leaving the opening, and cover the units on cool nights. In mild weather you can use Masonite on top instead of the tarpaulin. Add a little water nearly every day to prevent wilting but do not soak the containers.

Though even a few days of hardening will help, I recommend at least one week before setting the plants permanently in the Soil Bed. They can stay in the containers for a week or more depending on the weather, and when you find the right time to place them permanently into the Soil Bed as instructed in the next chapter.

*Warning:* Check plants each evening for wilting, especially on the first day, and if they are drooping, water them.

Set plants permanently in Soil Bed seven to fourteen days later, May 1 to May 5.

# 20•A Garbage Pail to Harden Plants

If you don't have the time to fuss with a greenhouse, use a garbage pail. The only thing that you will lose out on is a little earlier red tomatoes. But from then on all things will be equal, and although it is not intended to replace the plastic greenhouse I am happy to offer this helpful information. Plants hardened by this method will only be a little behind the greenhouse plants.

By setting the plant in a garbage can it is shielded from the winds (Photo 45), and you can either set the can in a shady spot or place a cover over it for a few days. It is advisable to cover the can at night. If it rains you should cover it. Also, make sure any water is drained from the bottom.

After a few days the plants will gradually become tougher and turn from light green to a deeper, almost olive color. It is important to water the plants almost every day since there is a tendency for them to wilt. They can remain in the garbage can a week, by which time they will develop into stronger plants. After a few days you can place them in the sun, but if they start to wilt, water them again and move them into the shade or cover them. However, you cannot keep the plants in the can too long. They cannot really get enough sun because of the high walls, and soon they will have to be put into the ground or they will become tall and spindly.

*45• Hardening plants in a garbage can*

But what if it is not the proper time to set out plants? Herein lies the difference between the garbage can and the plastic greenhouse. Using the greenhouse, plants can be set permanently in the Soil Bed even though it might appear that the weather is not suitable for tomato growing. But remember that the proper schedule for setting plants in the greenhouse and/or the garbage can must be adhered to.

## 21 • Seed Schedule Without Greenhouse

Start seeds indoors, March 15 to 20, a little earlier in a cooler house. Move plants outdoors for garbage can hardening (still in containers or pots) May 5 to 10.

One warm day in late April or early May is not a sign for you to rush out, buy plants and set them in the Soil Bed. Here are some unusual bad days in my area from my diary of 20 years:

May 2—30°. It ruined plants set out and damaged some covered lightly. (Not with my greenhouse.) It happened once in 20 years.

May 4—38°. Cold winds damaged unprotected hardened plants.

April 30. Light snow turning to rain, 32°.

Without the greenhouse you must be very careful, although you can always set up a makeshift protection.

You can also buy plants May 5 to 10 and harden them in a garbage can. But if you don't plan to harden the plants you buy, set them permanently in the Soil Bed on May 15. If the weather is extra warm, you can set them in a few days earlier. Hardened plants can be set permanently in the Soil Bed May 10, or a little later if the weather is bad.

The two separate schedules—greenhouse or garbage can—make allowances and are safe to use based on years of actual growing conditions. These schedules are for those living in the area of Valley Stream, Long Island, New York. New York City can be used as a check point although it is slightly warmer. Those living north or south of this area should make adjustments in the schedule.

If you live in the warmer parts of the country such as Florida and California that do not have cold winters and hot summers, you should not follow the schedules since hardening the plants as we know it is not important. I have been told that many gardeners in these areas use the summer as a resting period similar to our winters; they then get started again, probably in September. (Check with your local garden centers.)

Beginning users of my techniques should start a little later since having tall plants on your hands while it is still too cold outside can be a problem. However, those obeying the 68 degree temperature setting can start seeds about five days earlier than in the schedule. As mentioned earlier, seedlings kept in a cool room will set buds earlier, even though the plants might not look very stocky.

After the plants have been hardened for at least one week, they can be set permanently into the Soil Bed. However, if the weather is unsuitable then or you do not have the time you can postpone it a week or so later.

I usually do such important work on weekends when I have plenty of leisure. Indeed it is something that requires extra care, especially working in close quarters inside the plastic sides. The box unit is somewhat easier to work with, since you don't have to bend down. It is a little trickier working in ground units with plastic sides. But four plants are quite easy. Wait till you try eight early-variety plants in the same square!

Actually none of this is difficult, except that you want to

do a good job. You just can't toss plants into the ground and possibly damage them. After nursing these seedlings for two months, this is no time to snap one in half or tear the roots. If possible use an assistant. When you are ready to put a plant into the hole and you need help, it may be too late.

## 22• Transplanting Seedlings Permanently Into the Soil Bed

The plants in their containers should be just moist enough to handle. Too soggy is not good, and dry soil will simply fall apart. I usually wait until about three o'clock in the afternoon to make the transplant so I can head into the cool evening. Actually my plants, with a heavy set of roots, can never wilt (Photo 46).

Take the plants out of the greenhouse or garbage can and pick the best ones. If a wind is blowing, be very careful to put the plants where they will be protected.

Before actually placing the plants into the Soil Bed, dig up the soil a little or give it a good raking or fluffing. In other words, loosen it because it has been packed down with water.

*46• Milk container slit on four sides, ready for transplanting*

*Diagram 8•*

**Layout of plants in box or ground unit**

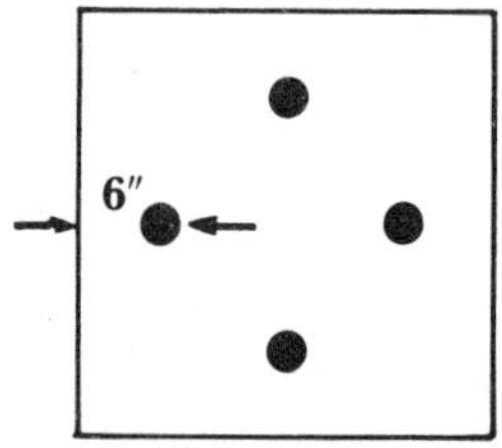


Notice in Diagram 8 that the plants are placed near a side, not a corner, both in the box and ground unit.

A tomato plant should be set a few inches deeper into the soil than it was in the container (see Diagram 9). For example, if the plant is 12 inches tall, allow 9 inches to protrude out of the soil. If 8 inches tall, allow 5 inches to remain above ground. You should estimate this so you will know how deep to make the hole. Then, using a trowel, dig the hole wide enough to take the entire root system without cramping. (You might have to place this soil in a pail temporarily to get it out of the way.)

Take one of the plants carefully and slit the milk container with a razor on the four corners (see Photo 46). I usually have a box or a little flat place where I can rest the plant when cutting away the container.

Carefully lay the entire root system in the hole (Photo 47). If it does not fit properly, remove it carefully and

*47• Moving plants into Soil Bed*

*Diagram 9•*

**Proper depth to set plants**

*48• Transplanted seedlings in a ground unit. Note plastic in place*

enlarge the hole. Don't be concerned if you cover a few of the branches. You can actually break them off. A strong top is all that is necessary. The plant will soon form new roots along its buried stalk.

When you are satisfied, push soil around the plant. Then with the palms of both hands, press heavily to firm the soil in place. This is very important to remove air pockets. When all the plants are in place, pour the leftover soil back in and spread it around evenly (Photo 48).

I do not water my plants or add starter solution immediately after planting. If you gave the soil a thorough soaking a week ago, it should still be damp enough and well supplied with nutrients. The plants should respond immediately.

## Guard for Cutworms

Before the day is over you must add a guard around your plants or risk having them cut down by the next morning (Diagram 10). These worms hide just below the

*Diagram 10•*

**Guard for cutworms**

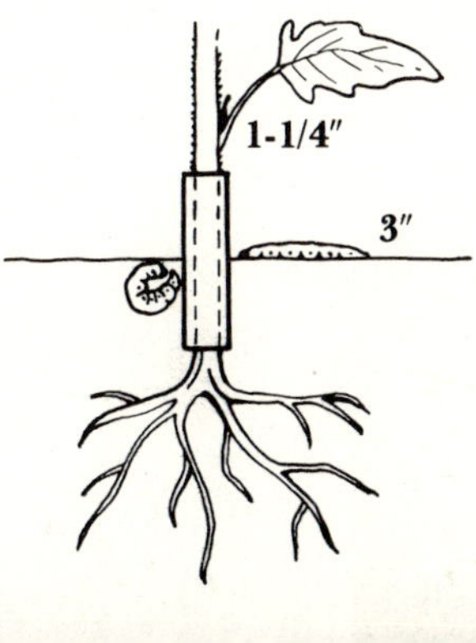

soil or under a stone, and you never know when they will strike. In effect you will be protecting the stem just above and below the soil line.

You can use the stiff cardboard from the milk container. I use a stiff piece of plastic that I have plenty of. Cut a piece about 3 by 4 inches. Have all the pieces ready. A roll of masking tape also helps.

With your finger, carefully remove the soil about 2 inches down from around each stem. If you uncover branches from the plant break them away, as mentioned before. Wrap the guard around the stem (the 3 inch measurement should be vertical), making a tube about 1-1/4 inch in diameter (Diagram 10). Add a piece of tape and push the dirt back around the tube. Once you collar all your plants, you are finished for the day.

The cutworm is not too intelligent, and he will eat a plant only at ground level. If the cutworms don't do any damage the first few nights they probably never will since they like only tender plants. However, unless you made the tube very tight around the plant, you can leave it in place for the entire season. The vines' girth will never exceed 1-1/4 inches.

## After Plants Are Set Out

In a few days you will add some water to your plants. Plants set out early in May don't want a wet and clammy environment. Tomato plants' likes and dislikes are similar to humans'. When it is hot they want to be cooled off, but when the air temperature is down around 50 degrees or lower, they want to be kept as warm and dry as possible. For this reason, I withhold water for a few days after transplanting. If the days are sunny immediately after, I'd advise a little water after a few days. If not, wait until the next weekend.

Set the nozzle of your hose for a fine spray and slowly let water trickle into the Soil Bed without disturbing the soil around the plants. If you haven't already got a good nozzle on your hose, get one—and not the trigger type. I've seen careless gardeners knock down a good seedling with a high-pressure blast from those gun-type nozzles. You should get a good brass nozzle that can be set for a fine spray. This will also be needed later for washing off tiny insects.

Since I have many units to water, I have two separate hoses in my garden. One has the brass adjustable nozzle for my vegetables while the other is only for watering the lawn. You can water your plants about once a week, unless it rains, then you don't water at all. After a week or so, you

*49• Two-foot bamboo sticks to train plants outward*

*50. A warm day, with top removed and plastic rolled down*

can use two-foot bamboo poles to separate the plants. Merely dig the stick in the center of the Soil Bed and carefully force the plants apart. I like to use two different sized bamboo sticks before switching to my permanent stakes. The two-footers can be used with the greenhouse covered. However, when the plastic sides are removed, I switch to three-foot sticks.

If you can't seem to hold the plants without tying, you should make a small bow tie. I use the green cord sold at garden centers. If they sell two different kinds, choose a ball of the heavier. This cord will also be used to tie the plants to the permanent stakes. But for now, merely make a loose knot so the plant will not move from the bamboo stick. It is only temporary.

You must adjust the sides of your greenhouse in accordance with the weather. This also applies to the Masonite on top and the tarpaulin that covers the unit. On a cold rainy day you can leave the Masonite on. At night, cover completely with the tarpaulin. On warm sunny days pull the plastic sides down to let more sun in. According to weather conditions, plastic down, plastic up; Masonite top on and off. The plastic sides can stay in place until late May. By that time you will be using the sides mostly to stop winds.

## Modified Version of the Greenhouse

Using the greenhouse can mean a few weeks of covering the plants at night and uncovering them in the morning, which, of course, means putting yourself out a little. You might want to try a combination of the garbage can and the uncovered greenhouse.

Harden your plants using the garbage can. Then set up the plastic sides as usual and set the plants permanently into the Soil Bed a little later (see schedule). Even though you don't cover the plants at night the fact that you set the plants out a little later, and the added protection from strong winds, will set you just a little behind the best early results, but far ahead of those gardeners using ordinary methods. And you can still cover the units should a real emergency arise. In a few weeks you will remove the plastic sides and your plants will already have little green tomatoes—that is, of course, if you followed directions correctly!

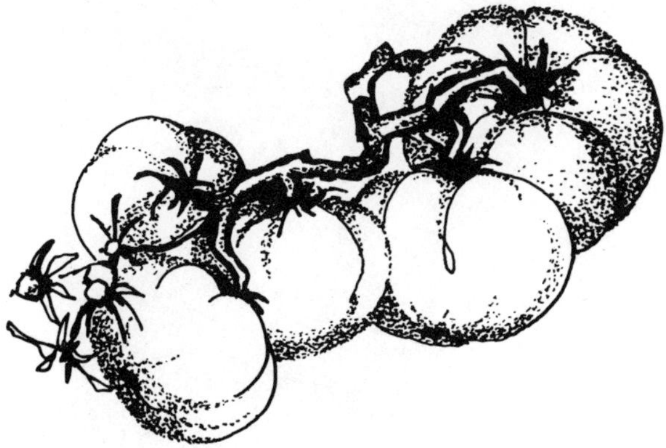

# 5 Basic Outdoor Care

## 23 • *Pruning to a Single Vine*

Once your plants are set into the Soil Bed they will make a rapid growth because of the deep, loose rich soil. In one week you can notice an amazing change. Watering should still be kept on a weekend basis.

Looking at Photo 51, you probably have noticed how the plants climb neatly up a stake, in contrast to the jungle of vines commonly seen in most gardens. I have often been asked, "Are those special plants that grow only a single vine?" The answer is, I prune my plants that way. This technique is used in commercial greenhouses, some research gardens, and by many commercial growers whose field produced tomatoes are intended for the high quality market, where an early crop will bring high prices. The

51• *Only one vine per stake*

plant is kept under perfect control, with a minimum of labor once set up, and easier insect control.

The single-vine method is a must for Multi-Plant gardening. It produces more tomatoes per square foot of space (you grow *up*, not *out*) and also the cleanest and earliest red tomatoes.

You have probably read that you get more tomatoes if you allow many vines to grow from the same plant. You will have to decide for yourself whether you want a jungle of vines and leaves or one vine loaded with large tomatoes.

The argument that you will get more tomatoes from many vines is true, but at the great expense of earliness, quality, and space. Why have many leaves and vines and few tomatoes, when you can have many tomatoes and few leaves and vines?

However, this technique of growing many plants in the same Soil Bed, can work successfully only if each of the four tomato plants is restricted to one stake. Once you have learned the simple technique of training each plant into one vine, you will be surprised how little work is involved in maintaining a unit.

Surely my photos showing loads of monstrous fruit should be proof enough. Greenhouses that must pay to

*52. Suckers emerge where a branch meets the main trunk. These suckers are almost big enough to remove*

heat every square inch of space want to make use of the upper part of the greenhouse, not only the lower. Hence they employ this method of pruning. Finally, insect control is much simpler, since the pests make a better target at eye level.

From the time the plants are set out in the Soil Bed, you must remove every sucker or new growth that appears between the main stem and a branch.

If you look at Photo 52, you will notice the so-called alternate or zig-zag way suckers appear, first right side, then left side. This is generally what to expect, but there may be exceptions. The point is, you must allow only one vine to climb the stake.

If your plants were about a foot tall when you set them into the Soil Bed, you will have seen these little suckers or growths starting to appear in a week. These must be removed, but you can wait until they are big enough to grasp, 1 to 1-1/2 inches. You don't need pruning shears. When the sucker gets to be about 1-1/2 inches long just break it off with thumb and forefinger. Think of the plant as a snake climbing a pole with one head and one body. Your plant must only have one head or growing tip. When you prune excess sprouts, all the plant's strength goes into

*53• Always prune below the uppermost flower cluster. Dark line shows the sucker that should come off*

fattening the tomatoes already set. Remember, in your units four heavy vines looking like garden hose are feeding those tomatoes, in contrast to one in an ordinary tomato bed.

Before you remove a sucker near the top of the plant, stop and think. You must not remove the snake's head, or growing tip in this case. In Photo 53, you see a sucker just below the cluster of buds and flowers. An excellent rule to follow is always prune *below* your upper flowers or buds. Anything above should be left alone. By chance you may allow one to get by until it's about 4 to 5 inches long. Better late than never—break it off!

As you look up and down your vine, especially near the bottom, you will no doubt see new plants growing off the main vine—pinch them off.

Warning: Do not remove large leaves or branches containing many leaves. Leaves are like solar cells from which the plant obtains much of its energy. If there are brown leaves on your plant, removing them won't benefit the plant. When they are completely dried out, however, they can be removed for appearance and possibly sanitary reasons.

## 24 • First Buds, First Flowers, and Pollination

Once the plants are set out, keep your eyes on the buds and flowers, for it is from these that the first tomatoes will emerge. The accompanying close-up photographs explain the steps from the first tiny bud to the setting of an entire cluster of tomatoes.

54• *First buds*

55• *First tomato flower fully open*

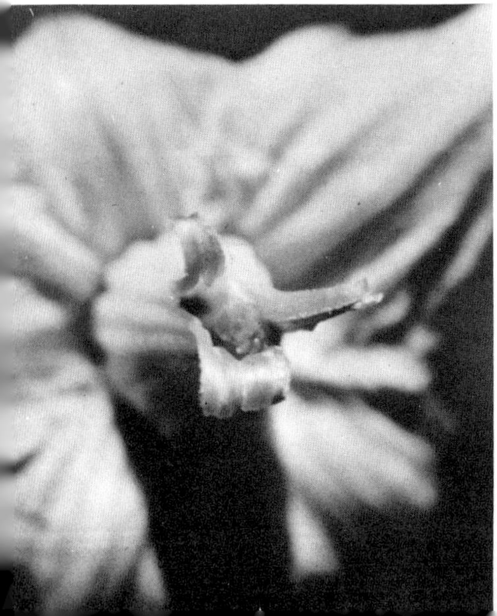

56• *Three curved structures are the pollen-bearing anthers. Directly between them, only a fraction of an inch away, is the stigma*

57• *Side view of a similar flower, with petals and anthers removed. The stigma is at the top of the long pistil tube, at the bottom of which is the ovary—your future tomato*

*58. First tomato breaking through, with dried blossom still adhering*

In Photo 54 you see the first bud of the season magnified about five times. There is little doubt that the first red tomatoes will come from one of these. In only a matter of a week or so, those tiny buds will open to a yellow flower or blossom (Photo 55). It takes about a week after the flower opens until a small tomato is visible. In this cool environment, it will be about 45 days between the time you see the tiny tomato and it is ready to eat.

They say you can't sell a book without sex in it, so Photos 56 and 57 show close-ups of the plant's sexual parts. In Photo 58, appearing about the size of an average pearl, the blossom is almost dried out as the tiny tomato appears. In a very rare shot you see the pistil dried out and still hanging on, indicating a dry and wind-free week.

The subject of pollination is foreign to many inexperienced tomato growers. They couldn't care less how, why, or what pollination is—"Just show me a tomato."

Many gardeners take pollination for granted. They assume that a tomato automatically must set, once the flower dries out. Yet this is not so; there must be a transfer of pollen first.

## Bees Not Needed

Contrary to popular belief, bees are not really needed to transfer pollen on a tomato flower however valuable they are to other plants such as cucumbers and all the vine crops.

In order to have production of any seed, pollen must be transferred from the tips of the (male) stamens to the stigma atop the pistil tube (female). The tomato has what is known as a perfect flower—it has both these male and female organs in the same blossom. By contrast, a cucumber plant has two separate flowers on the same plant. This means pollen must go from one cucumber flower to another, which must be done by a bee, another insect, or by man using a brush.

In the tomato flower it is much simpler; the pollen must only travel a fraction of an inch, and the wind usually accomplishes this.

Looking at the close-up Photo 56, we see a head-on shot into the stigma, which looks like a wart sitting on top of the pistil tube (Photo 57). Surrounding the pistil, looking like an Indian teepee open on top, are the yellow stamens. In the other picture the stamens were cut away to reveal the pistil, and the tiny tomato at the base of it. Here blown up many times original size, the small tomato is actually only about 1/10 inch in diameter.

At the tips or top of the stamens are the anthers, the pollen-producing portions of the flower. Almost within touching distance, the flat topped stigma sits waiting, with a sticky material upon its surface to catch and hold the pollen. When the flower is fully open, pollen bursts forth and almost any movement of the plant will cause this dust-like pollen to fly around. In the greenhouse, where there isn't any wind, vibrators are used to shake the plant causing the pollen to fly around.

Once contact is made, a reaction is set off which works

downward to the waiting ovum or tiny tomato. Here fertilization takes place, and the tomato soon starts to swell (Photo 57).

## Self-Pollination and Cross-Pollination

The above explanation is an example of self-pollination. However, if pollen is brought from a different plant by a bee or man, this is cross-pollination, and the resultant cross will produce hybrid seed as explained earlier.

There are specialists that do this type of work, developing new and better plants; these people are called hybridizers. But how can they know for sure whether self- or cross-pollination had taken place? They have a sure way. Just before the flower is ripe or opens, the stamens are cut away, just as I have done in the photo, and covered with a bag. Then, at the proper time, pollen is brought from the donor plant by brush and placed on the stigma of the recipient. The flower is immediately covered again to exclude all other pollen. In a few days the bag is removed. This tomato is allowed to ripen fully and the hybrid seeds carefully processed.

Temperatures well above 90 degrees can be a problem. However, it is the night temperature that is most important; 80 degrees seems to be the cut-off for setting tomatoes. In other words, if your night temperature does not drop to at least 80 degrees or less, your flowers will probably dry up, and no small tomatoes will emerge.

I have heard of using partial shade overhead and cool water to hose the plants at night. But these techniques are for extremely high temperatures only and should not be used under normal conditions. Ideal temperatures for growing tomatoes appear to be 80 to 85 degrees in the day and 60 to 65 degrees at night.

Since I work so far ahead of the season, thanks to my

plastic greenhouse, my flowers are out when the air temperature is still very cool. If the weather is cool as it usually is, some of the flowers may drop off without setting a tomato. I have found something that will help set the tomatoes during these cool days.

## 25• A Liquid Spray to Help Set Early Tomatoes

For the past twenty-five years I have sprayed my first blossoms with a liquid that is supposed to help set those first blossoms and produce early tomatoes. This does not mean that tomatoes will not set without its use. But I have the feeling that the use of this liquid produces a better first setting, so important to tomato growers.

Like organic gardeners I am opposed to using things that are not natural, but I must try to produce as high a yield as I possibly can. I use this liquid only during the cool days and discontinue it by late May. I have read that the amount of liquid used to effect the necessary reaction is about the size of a single pollen grain, which must be viewed with a microscope to see. It seems to me, therefore, that the risk of any serious contamination is unlikely.

### Many Products to Use

To help set tomatoes many products are available —liquids, powders, and aerosol spray. I prefer the powder which is made into a liquid. You can use an eyedropper to spray the liquid into the flower. However, I have designed something to make things easier that does away with aerosol cans, eye droppers, and other ineffective means of applying this liquid. This special applicator is made from

*59• Equipment for blossom-spray application*

an auto battery syringe, a length of rubber tubing, and the glass portion of an eyedropper (Photo 59).

I have also developed my own methods for making applications—these are not the same as the directions call for, but I get good results.

When the flower opens (Photo 60), I spray a little liquid into it. It will have an effect only when the flower is fully open, but you can start a day before and continue a day later to make sure. There is no harm in extra applications. Each day, go from flower to flower. I usually make my applications before going to work, but midday or afternoon is an ideal time to do it. The applicator makes this operation quite simple because I can place the tip of glass with one hand just outside the flower and give the bulb a squeeze with the other hand.

*60• Only fully open flowers should be sprayed*

## Produces Seedless Tomatoes

According to the theory that I have read, the pollen, which is supposed to be the male contribution to fertilization, is completely by-passed when using this liquid—(almost like making a baby without daddy around). This liquid activates the female part as in most normal reactions, and a small tomato is formed—but without seeds. It is a sheer delight to cut open a seedless tomato. It looks like a piece of rare roast beef! What looks like tiny seeds is really the structure that would hold the seed. In some cases you will see a few seeds present. This probably indicates that the liquid and some natural pollen both arrived at the same time.

If you are an organic gardener, you can disregard this chapter on a liquid spray to set early tomatoes. Since it is effective only for a few weeks during the cool days of spring, and you can certainly continue to have good results without it.

A good healthy plant growing under optimum conditions will always have flowers and tomatoes setting (Photo 61). When the plants are loaded with tomatoes there may

*61• A healthy flower cluster ready for spraying*

*62• A perfect setting of tomatoes*

be a hesitation in the growth of the top buds. Also heat waves (above 90° F.) and long periods without sun will also prevent the set of new fruit up top. But by and large a good plant with many clusters, containing many tomatoes per cluster, is a sure indication that things are right (Photo 62).

## 26 • Mulch

Mulch is just about anything used to cover the soil near a plant so the hot sun won't dry out the roots. A good mulch is important in all types of gardening, and to anyone using the Multi-Plant Method, it is a must.

As soon as it starts to get warm you have a double purpose for adding this protection in our units. Although mulch is used to smother weeds and conserve water, your main use of mulch is to shield the soil and the network of roots just below it. Not only are you concerned about keeping the hot rays of the sun off the Soil Bed, but you also need some material to act as a shield when you water. Concentrating the hose in one spot can easily wash away the soil and expose the roots. Working with a small 20 by 20 inch area, it is imperative that the roots of the plant never be exposed. If they are they will dry out causing Blossom-end Rot.

As I said, just about anything can be used as a mulch.

*63• Bark mulch in place on Newspaper Box unit*

But I have done research on what is best for our units, and I have narrowed my choice down to two materials. My first choice is a few inches of medium size bark (not pine bark), available at garden centers. If the medium size is not available, the large size may be used (Photo 63).

Without doubt, a few inches of bark is best. A little extra is better than less. But do not buy a bag marked pine bark; it may be too acid.

Bark is absolutely great to work with, and in our units it will not mat down or deteriorate as other materials do. I can direct my hose at a few inches of it and all the water slowly penetrates downward without disturbing the mulch or the roots. These then, stay moist all day long. And if carefully removed at the end of the season, the mulch can be used again the following year.

For those who do not want to invest in a bag of bark, dried grass, weeds, or hay is acceptable, though not as good. A few inches of this loose material works very well, but it loses its effectiveness as it starts to deteriorate and mat down. By that time, the season may be almost over and it may not matter much, especially as you approach the cool days. But if you are concerned with good looks, the reddish bark will do it.

## Wait Until It Gets Warm

Mulch should not be applied until there is a need for it. In the cool days of May, it is an advantage to allow the sun to warm the soil. But as the temperature during the day rises to 80 degrees and above, it is time to add mulch since you will also be watering more often.

Since it is almost impossible to make sure that every root is completely covered, even when bark is used, it is still possible to get an occasional black spot. But you will have far less problems from Blossom-end Rot than if no mulch or an ineffective one is used.

### For Organic Gardeners

If using unprocessed manure with pieces of straw, etc., you may or may not consider this as your complete mulch. Surely the fine dried material would not be enough protection. Thus I still recommend a few inches of medium size bark on top of the manure as protection from the hose water.

## 27 • When and How to Use Permanent Stakes

Friends who still grow tomatoes the old-fashioned way spend many evenings tying plants. It seems that there are always plants to tie, and an endless amount of sticks are needed. But come a good windstorm, all the plants are on the ground, a dozen or more large tomatoes all over the place, and the vines snapped in two. This has never happened to me, and it would take quite a storm to do it.

Earlier, I explained how to prune your plants to a single vine. This chapter will explain how to make a stake that will last almost forever and can nearly withstand a hurricane.

I adopted the two-piece stakes about twenty years ago; they replaced pieces of broomstick, bamboo, and many other things that were not only ineffective but also not uniform. Once the plants have reached the top of three-foot bamboo sticks, it is time to make the transfer to the permanent stakes.

The stakes I use and that are shown in the photo are made from expensive lumber. However, compared to the aggravation you will have by trying to cut corners or save money, it will really pay you to do it right the first time. One gardener wrote me that the tall stakes are a great idea

*64• Two-part stakes, notched at the bottom for easier insertion into the Soil Bed*

to keep the slugs off. I assume he means that slugs usually eat tomatoes that are on or near the ground.

## Stakes for Ground Units

The ground units require a two-piece stake (see Photo 64). These have a few advantages. If the base of the stake in the ground decays, it is easy and inexpensive to replace. Another advantage of the two-piece stake is that it is extra long and very sturdy.

The lumber I use is called 1-1/8 inch square clear pine. It is actually about 1 inch square. For the ground units you must order four pieces of the 1-1/8 inch square, each six feet long. In addition, you should order four pieces of inexpensive furring strip, three feet long. (It is important that these short pieces have no knots in them, or they will be useless.) Using a hand saw, cut into the end and snap off a piece on both sides to form a point, though a point is not absolutely necessary. (For the box, order four pieces 1-1/8 inch square and seven feet long.)

The two pieces are made to overlap by about one foot and two 1/4 inch holes are drilled through both. They are held together with nuts and screws. They will be eight feet long.

All the holes used in my units to hold stakes are exactly the same, so all stakes are interchangeable. I made a template or guide from a flat piece of steel as described in Chapter 6, to show where the two holes should go. Merely clamp this onto your stake or on the side of the box, where screws will be set in, and drill two holes. The two 1/4 inch holes are about seven inches apart, and a few inches in from the end of the stake.

The seven-foot stakes for the boxes will need the same two holes. Holes for the box stakes should be made a little larger to afford a little extra clearance. You can use a larger bit (5/16 or 3/8 inch) or enlarge the hole using the 1/4 inch drill.

## Add Wood Preservative and Paint

Once the holes are drilled, give all the lumber a coat of wood preservative. Let it dry, then give it a good coat of an outdoor enamel paint. I buy one quart of a dark green color (see boxes, Chapter 6).

Just before using stakes, you must join the two pieces together using 1/4 inch by 20 round-head screws, 2-1/2 inches long.

Place the two-piece stakes on ground units just outside the pieces of Masonite of the 20 by 20 inch square hole.

I have been asked how to hammer an eight-foot stake into the ground. It's quite simple. First make a pilot hole using a piece of pointed furring strip, then place the stake in that hole. Depending on your soil, the hole should go 12 to 18 inches into the ground. It should go in at a slight angle away from the center of the unit so that the plants will spread out a little up top. Once the hole is made, the stake is forced down into it a few times and soil packed around it.

*65• How the stake is made to angle outward instead of straight up: use a nut as a stop on the upper bolt*

## Adding Stakes to the Box

As mentioned earlier, four stakes must be attached to the sides of the box using two preset screws on each side. The stakes with predrilled holes are put in place and secured with washers and nuts. When fastening the stakes onto the box, they should slant outward the same as with the ground units. This is accomplished by putting a nut as a stop on the upper screw (see Photo 65). If you place a nut about 1/2 inches out from the box, the stake will never be able to lie flat against the box. It will slant out as it goes up, allowing the plants more room. For this reason, the holes on the stakes must have extra clearance.

## Maintaining the Stakes

If you try to cut corners and use flimsy stakes to hold your plants, remember that I told you so when you find your plants and tomatoes smashed on the ground after a storm. Even if they're left outdoors all winter, these good stakes can go for a few years without repainting. However, the piece that goes into the ground must be watched carefully each year for termite attack. If only the paint is scratched off, it can be repainted, but if there are bad holes the piece must be replaced.

# 28•Tying Stakes to Plants

Pieces of white bedsheet and nylon stockings may have been good enough years ago, but today they make for a sloppy garden. Pick up any book on vegetable growing and you probably will still find these old slip-shod ways suggested.

In keeping with my neat way of gardening, I use a ball of green cord to tie the vines to the stakes. Most garden centers carry this cord, and you should choose the heavier type if you have a choice.

It is important to tie the vine as close to the stake as possible. If it's too far away, it can sway back and forth in a strong wind, causing damage. Actually, you really don't have to hold the plant up, merely keep it from swaying.

I have found that a simple double knot gives good results and eliminates the extra work of first tying the cord to the stake, then to the vine. In other words, all you do is lay the vine against the stake and make a fairly loose double knot (see Photos 66 and 67). A nice close tie will withstand

66• *The right place to tie your plants—just below a branch, but above a cluster*

67• *Tying plants to bamboo extensions (see Chapter 38). But at least this much length of vine should be allowed to swing free*

very high winds because the stake sways but not the plant. Once the vine finds a base, it will not usually slide down any more, and all you have to do is make a tie every 8 inches or so, as the vine climbs up the stake by itself.

My plants have gone up to 10 feet and more without any problems of slipping down or swaying. Thus, it makes a lot of sense to use this simple tie, rather than the double tie recommended in many garden books, which is not only time-consuming, but requires patience and nimble fingers, to say nothing of working close to sticky tomato plants on a hot summer day! However, it's still a little tricky. You must make sure you don't damage the buds or tiny tomatoes. If your tomatoes are squashed against the stake after you finish a tie, you are doing something very wrong. Look at the photo of me standing next to my box and you will notice how all the tomatoes hang freely.

## Tie Above a Cluster of Tomatoes

It is important that you make a tie above a cluster. This way, as the plant sags under the weight of the load of tomatoes, there is no chance of damaging the feeding stem of the cluster.

You must also make sure that these clusters are always facing *away* from the stake. What seems to complicate things a bit is the fact that one cluster will be on one side of the vine and the next will be on the other. When you are ready to make your next tie, you might be squeezing the topmost tiny buds and flowers against the stake. When this happens, I usually untie the lower piece of cord so I can swing the vine freely and place it properly against the stake. Never try to force the vine into position, or it can snap like a string bean. Never try to tie right to the top. Always allow a few inches to swing free. You don't need to tie again until you see a fairly long length of vine swinging freely (Photo 67). But don't let it get too long, or it will snap the next time you have a strong wind.

### First Aid for Broken Vines

If you ever do snap a vine while trying to maneuver it into place, quickly put it back in its original position. Try to get someone to help you as you tape the vine to the stake above and below the break and finally cover the break with tape. If the upper part of your plant does not wilt by morning, your operation was a success. Of course, the upper part will slow down somewhat, but it is much better than if completely broken off. After a few weeks, remove the tape from the stake but keep some wrapped around the old break.

## 29 • Principles of Fertilizing and Watering

Since four plants are all crowded into one small Soil Bed, it makes sense that the plants will need frequent waterings. And of course the more often water is applied, the more often your supply of plant food will have to be replenished. Thus, once the plants are loaded with tomatoes and the weather gets warm, I establish a daily watering and weekly feeding schedule—and it's almost foolproof.

As one old-timer said, "Such a thing is impossible. I've seen tomato plants twisted up like corkscrews because someone carelessly applied too much fertilizer. And I've seen starved, light-colored plants with marble-sized tomatoes on them. Feeding is a real art."

*68• Newspaper Box being fertilized from a gallon jug of concentrate*

The technique employed in feeding my plants might be called a "buffer system." As soon as there is danger of a toxic condition caused by too much fertilizer, water is brought into play to act as a neutralizer or stabilizer. But this happens automatically during the watering and feeding process so that you aren't really aware of it.

Photo 68 shows a unit of four plants. You can see how easily they are watered. Compare this high-efficiency type of watering to the farmer or home gardener who uses a sprinkler to irrigate his plants. I find it shocking that a sprinkler wastes from 180 to many hundreds of gallons of precious drinking water per hour. Since most sprinklers have a set circular pattern, some of the water falls on places where plants don't even grow. And farmers will water in between the rows, since there is usually no way to avoid this waste.

No doubt many roots will draw more water, but much of the water is wasted anyway either by evaporation from the surface or by water working its way deep into the subsoil. But the water from a sprinkler only penetrates a small amount in one hour, so it is extremely difficult to get water down deep enough to encourage long roots that make for big healthy plants.

Water is not only held to a minimum with my system, but it is applied in a very effective and efficient manner, penetrating downward rapidly, insuring a deep, heavy root system. The units, being confined, get an 18 inch penetration in minutes.

It is my feeling that the box unit is a perfect example of water efficiency at its best, at least as far as the home gardener is concerned. The four gallons added daily are almost all used by the four giant plants. The many large leaves and some eighty or more tomatoes that are always hanging will surely drink up the daily supply of water on a nice hot day, which is almost every day during the summer season. The fact that the plants will wilt in a box unit if not watered each day is proof that all the water is used up—meaning extremely high water efficiency.

Frequent waterings also supply large amounts of needed oxygen to the root system. I even refer to this condition as "super-saturated." As the water gravitates downward, it forces air bubbles through the roots. As these tiny bubbles break, they release this life-giving oxygen to the plant. This is one reason why the plants maintain such high production over such long periods.

Keeping my plants well-fed means feeding them to almost dangerously high levels. But they do not remain long under these conditions when water is soon applied to relieve this danger. The important thing is that it really *works*, whether you are a pro or a beginner.

## 30 • Preparing Liquid Fertilizer

Very shortly you will need to start feeding the tomatoes that are now on your vines. Many gardeners think there are only three chemical elements in a bag of fertilizer: nitrogen, phosphorus, and potassium, since these must be listed on each bag by law, showing how much of each is available in the bag. These three chemicals have a great influence on the growth of plants and so are most often spoken of. However, based on material published about twenty-five or more years ago, it was found that eleven different chemical elements were needed to sustain good growth. Six of these were supplied in large amounts, whereas only traces of the other five were needed, hence they were called trace elements. The six major elements are: nitrogen, phosphorus, potassium, calcium, magnesium, and sulfur. The five trace elements are: iron, zinc, copper, boron, and manganese. In the last few years I think I have read that molybdenum is also a trace element. All eleven chemicals must be present in exactly the right proportions in order for the plant to grow properly.

In agricultural universities studies are made using water culture in individual jars or bottles. The solutions are made by using very pure chemicals, weighed out as accurately as any prescription used for people, usually using distilled water. During my visits to one agricultural college, I have seen some twenty-three jars side by side, all growing tomato plants. One would wonder why so many jars should be needed. In one group marked *deficiency*, each plant had one element missing. In another group marked *toxicity*, each plant had too much of one of the eleven chemical elements. One jar called *control* had all chemicals in the proper proportions. Students studied the symptoms caused by the various deficienceis and toxic conditions. They could then go out in the field and tell what was wrong with a plant and how to correct it.

It does not matter whether you are using all organic plant food or a bag of commercially prepared fertilizer. If there is a deficiency or too much (toxicity) of any one element the plant will develop abnormal symptoms.

In late June your tomatoes are large enough to be picked. Now all you have to do is get them to turn red. It has been known for years that there is a definite relation between seed formation and phosphorus. Feeding the plants with a formula high in phosphorus, such as 5-10-5, will hasten the ripening process. Then, if you follow carefully the directions on when to start feeding your plants, you will have early red tomatoes.

Our plants very rarely have nutrition problems since my feeding is almost foolproof, as mentioned, done with almost the same accuracy of a physician administering a hypodermic injection. The feedings are started at a low concentration and brought up to full strength at the proper time in the schedule. It is almost impossible to apply an overdose, while at the same time applying maximum dosages for optimum growth and production.

It is best to prepare in advance for this next important move. You have purchased a bag of 5-10-5 fertilizer from your garden center. I recommended the 50-pound bag as the most economical, but there are smaller bags available.

You should not substitute fertilizers; my system of feedings took years to perfect. This is not to say that 5-10-10 or 10-6-4 will not work, but try my way first before you start to experiment.

As mentioned before, I have eliminated the use of manure in my soil mix. As this precious stuff started to become scarce prices started to go up, and you had to go far and wide to get the real thing. Not wishing to be dependent on a scarce item I gradually worked 5-10-5, a complete fertilizer, into my soil mix. I have since developed a technique where I use 5-10-5 in both dry and liquid states with perfect control.

Usually, fertilizer is spread on the ground around the plant when feeding—not nearly as effective as my method. I make a liquid from the dry fertilizer by putting it into a sort of tea bag and allowing it to soak in a pail of water. This holds back the solid material which can cause problems if dumped into the Soil Bed. I suppose it is only a matter of time before I use those easy-dissolving types of fertilizers, but I have not yet tried them. Mine is more economical and can be used in the original Soil Bed preparation also, but I must admit it does require a little extra work.

## Pails, Bottles and Pieces of Cloth Needed

You should have ready an ample supply of pieces of cloth, one-foot square. A finer material is preferred to a heavy one which is difficult for liquid to drain through.

Weigh out 1/2 pound of 5-10-5 and dump it in the center of the cloth. To simplify things, you can use the bottom of a quart milk container to measure half a pound. Cut off and keep the bottom 2 inches of the container. Measure up 1-5/8 inches and draw a black line (Photo 69). Fill the container to this line and you have approximately one half pound—enough for your use. This amount will make two gallons of concentrate. Tie the bag with a double knot, making sure no dry material will get out. It is best to use a piece of cord that will not soon deteriorate, since you can use both the cloth and cord a few times more.

*69• Equipment for making liquid fertilizer–measuring container, cord, and 1 foot square cloth, plus 1/2 pound of 5-10-5*

*70• Tie 5-10-5 fertilizer in bag and agitate in a ten quart pail*

Then put about two gallons of water in a ten-quart plastic pail, and swish the bag in it for a few minutes (Photo 70). You will soon see the liquid become cloudy. Do not try to dissolve all the material in the bag since about 80 percent is only crushed stone. Drop the bag into the pail of water and set aside for a few days to slowly dissolve out more of the nutrients.

Have ready two one-gallon jugs or bottles, preferably plastic. A funnel will also help. Mix the contents of the pail well by swishing the bag around and pour into the two bottles. To better distribute the liquid, fill one bottle partially, then the other, until both are filled. If you are short, add water.

## What to Do With the Used Material in the Bag

The bags are used only once and then set aside to dry out. They are then opened carefully, the contents dumped onto newspapers, folded into a package, taped, and placed in the garbage pail. The cloth can be used again until holes develop in it. You might be able to use the crushed stone as drainage material, or perhaps for other garden use. It still contains some fertilizer and you might add it around some of your flowers.

In order to save time, you can prepare many bottles in advance and cap them. A few minutes on some dull evening will keep you well supplied. *Don't* use the bag of fertilizer again, even if it makes the water cloudy; you will only be fooling yourself. Although not completely used up, old fertilizer is not satisfactory—tomatoes will take longer to ripen if you use it.

For organic gardening, skip the 5-10-5. But when the plants have set tomatoes, it is time to think about feedings. When the green tomatoes are about 1 inch in size you should cover the Soil Bed with manure under your mulch (see Chapter 26), and also sprinkle in some bone meal or other phosphorus-rich material. Above ground, manure can rarely cause a toxic problem as it can below.

Because we use such a concentrated Soil Bed, daily watering is recommended, especially in the box. (See watering and feeding directions telling when to start.) However, you will not need to add any more fertilizer. But consider these alternatives: you can add more manure after a month or use liquid manure weekly just as I use 5-10-5. Organic gardeners of years ago know of the benefits of this liquid, made by filling a sack with manure and tying it so the material can't get out. It is then dumped into a barrel of water and allowed to soak. A gallon of this brown liquid is poured into the Soil Bed. When more is needed, usually only water is added to the barrel. Add liquid manure every week and try to get some additional phosphorus. But even if you skip feedings, plants will probably still do well.

## 31 • Osmosis: How the Plant Takes in Food and Water

Have you ever fertilized your lawn and accidentally let a load of the stuff fall in one spot? Of course, the next day or so produced the most beautiful section of brown straw right in the middle of the dark green grass.

Do you know the true explanation of how that patch of grass turned brown so fast? Blame it on osmosis.

Do you recall the days when you studied high school biology and had a drawing of what looked like a finger, possibly hanging on the wall? This was an enlarged view of a root hair. These tiny hairs line the roots and take in liquids needed by the plant, almost like the human mouth.

The skin or covering of the root hair is referred to as a semi-permeable membrane. Liquids can flow through this membrane only if there is a difference in the concentration of the two liquids on either side. However, the whole point of osmosis is to neutralize the liquids on either side. If there is a strong solution on one side and plain water on the other, the weaker solution will flow into the stronger.

Fertilizer, whether it came from a bag of 5-10-5 or from manure, can affect the osmotic flow or equilibrium. Normally a little fertilizer around the grass roots results in a slow but steady flow into the plant, where the concentration of minerals is stronger. If we increase the concentration of fertilizer, liquid flow will gradually slow down. But if we really over-fertilize by dropping a large amount of chemical in one spot, it creates a strong solution outside the plant. This creates a reverse flow and actually dries out the plant, resulting in straw-colored grass.

## Back to Your Tomato Plants

Commercial gardeners have always tried to keep the concentration of fertilizer as high as possible without causing a toxic condition. Thus production is kept very high. As the concentration of fertilizer increases, the flow of water into the plant will keep getting slower, but it will not stop completely. Such a high concentration will produce the highest yield, but it is quite risky and can only be done by experts.

In the next chapter on feeding, I feel I have come close to reaching the maximum. And yet, if the simple directions are followed, your plants will never be in trouble because of too much fertilizer.

# 32 • Daily Watering and Weekly Feeding

Exactly when to start feeding is one of the important parts of my method and any other method of gardening.

It must be done before the plant develops a serious diet deficiency. Yet extra food supplied prematurely can be disastrous, resulting in twisted and distorted plants, with wholesale loss and dropping of the potential tomato-producing flowers.

Everything works by schedule, almost to a preset date on a calendar. Thus, if you set out healthy plants as I do, early in May, by early June your green tomatoes will be about 1-1/2 inches in size and you should start feeding. As a matter of fact you can forget the date and go by the 1-1/2 inch tomato size.

Since watering and feeding are closely linked, it is important that the special watering and feeding schedule begin together. Of course, you have been watering your plants a little right along; in most cases, once a week was enough. But now the weather is warmer, and the tomatoes need additional food if they are to develop into the large ones you look forward to seeing.

## Have Liquid Fertilizer Ready in Gallon Bottles

Preferably on a weekend when you don't have to go to work, you will apply your first feeding, using only one half concentrate for about two weeks. Take one of the gallon bottles you prepared. Take half the contents and fill with water. These two diluted gallons will be used for the next two weeks.

Preferably in the morning first water your Soil Bed, adjusting your hose to a medium spray. Make sure your mulch is in place. Approximately two gallons are applied first, allowed to settle down, and two more gallons added. You will soon get to know how much two separate applications are when using a hose, so that you won't have to measure it in gallons. It will be more like half a minute or so at a time, depending on the size of hose and the water pressure. Three gallons would be enough while the green tomatoes are not full size, but with good drainage a little too much water is not a problem. Four gallons is about the most you should add at one time.

Allow the water to settle down for about fifteen minutes and add the gallon of liquid fertilizer, spreading it all around the Soil Bed. Repeat this process a week later, but water your unit every day.

In two more weeks your green tomatoes should be much

bigger, and you should start your schedule of full concentrate each week. This will continue until October.

Even if you continued on one-half concentrate all season, you would still get good results, and you would probably never know the difference. However, tests made many years ago showed me that the heavier concentration produced larger and earlier red tomatoes and set bigger clusters up top, while the one-half concentrate-fed plants had a tendency to drop flowers once the plants were loaded. The new growth was also not as good and ripening was somewhat delayed.

All good things must have some disadvantages and the box has one big one—it should not go without water for more than twenty-four hours when the plants are loaded with large tomatoes and the weather is hot. If you omit daily waterings the ground unit will not get into trouble as fast as the box. But with some eighty large tomatoes hanging on four giant plants in 90 degree weather, the box will probably wilt if you don't water every twenty-four hours.

## Do's and Don'ts

*Don't* water your unit all day long. You will wash out the nutrients ahead of time. Even if it is a hot day, the units are designed to go at least twenty-four hours without wilting.

*Do* use a nozzle on your hose that can be adjusted; not the trigger type which can cause damage if the water is allowed to come out full force.

*Do* give your Soil Bed a sprinkling after a very hot day. This will keep the surface roots from drying out. When in doubt if it is a hot enough day, sprinkle it anyway.

*Do* feed the same time each week if possible.

When you are growing in boxes on balconies, you will be watering with a pail or gallon jug instead of a hose. Try not to dig up the soil as you dump in the water. Here's where a few inches of medium-sized bark is so important (see mulch). Once the plants are loaded with tomatoes and the weather gets hot, about four gallons of water will be used each day.

Slowly and carefully add two gallons and allow to drain before adding the other two. Spread the water all around, not just in the middle of the box. Your regular watering is best done in the early morning before going to work. On a very hot day you can add some water about five to seven o'clock in the evening to moisten surface roots.

### Do I Water in the Rain?

The condition of the box is most critical during the height of the season when the plants are under heavy strain. Let's be realistic about it; can a rain add three to four gallons of water to the Soil Bed? If, in your opinion, it did, then don't add water. After lighter rain, add less than the normal amount if you wish. It's not critical, and extra water won't hurt.

When you approach late September and early October you will have to use your own judgment about watering, but even at the worst you should not be too concerned since the season is now behind you, and you have little to lose. But the rule is to water less during cool days. You can stop or ease up on your feedings.

## 33 • Controlling Tiny Sucking Insects

Organic gardeners will be glad to learn of my simple ways of controlling insects, which don't include the use of poisons. Rather than try to cover all insects in one long confusing chapter, I prefer to discuss problems as they appear during the growing season. In an earlier chapter, I explained how to keep the cutworm away from your tender plants with a tube of cardboard or plastic. This chapter covers tiny sucking insects.

As soon as the weather starts to get warm, usually by early June, aphids and other small bugs make their appearance. As with most insects, they have a definite life cycle so you can expect to see them leave by early July. The most common of these tiny sucking insects come in three different colors, red, white, and green.

It is interesting to note that while one plant will have a heavy concentration of insects, another tomato plant in another part of the garden, will not have a single one. And in some years there will be many, while in others almost none. But it is a known fact that insects will attack the unhealthy plants first. You should not have many insect problems high on a balcony. However, after a few years you might, since insects tend to hide and live in the soil you will be bringing in.

I don't try to destroy every bug that lands on my plants. Good control simply means containing the insects to a point where they will not do serious damage. A few holes in a leaf here or there is really nothing to get excited about. I also go along with our organic friends and refuse to spill poison on my plants. As for the insects, you can find them under a leaf, or elsewhere, but mostly they are attracted to the new top growth where the plant's skin is thinnest, and to the yellow flowers or blossoms. A certain amount of caution must be observed so as not to damage these flowers that your potential tomatoes must come from.

There are two ways to control these insects without sprays and powders: by importing ladybugs and praying mantises to eat them, or by doing what I do—annoy them with a fine spray of water. Pure water sprayed on these tiny insects has an adverse effect on them. While they suck the juices from the plant, they do not like to be disturbed. Water will not kill them, but it really annoys them and tends to dilute the sap they live on, with the result that they usually do little damage. The water spray must be applied daily and sometimes twice a day.

It is most important that you use a nozzle on your hose that can be adjusted to a fine spray. Too much pressure can damage the flowers. Start by directing a fine spray at a concentration of insects. Actually, you are not trying to get every one off, but no doubt you will lessen their numbers since many will be washed away. Again, be very careful near the flowers; soaking might actually prevent fruit set. Better to retain a few insects than lose good blossoms.

Don't be surprised to find the tiny critters back the next day. It's a case of who has more patience. Under unusually bad conditions you might have to spray them in the morning and again in the early evening. But soon you will notice that they will gradually disappear without causing much damage.

If you had a heavy concentration they might leave some of your leaves dried and a little distorted. Don't be too concerned; soon new growth will take over. At worst, you may have a few deformed or imperfect tomatoes.

# 6 The Hot Days of Midsummer

## 34 • Insulating the Box and Hosing the Plants

It is a decided advantage to have the sun heat the box during cool days so the soil and roots can warm up. But about early June or possibly sooner, depending on the temperature, you will need to insulate the Newspaper Box from the hot rays of the sun.

Newspapers should be wrapped around the box and covered with clear plastic. Actually, the plastic covering is added first, forming pockets, then filled with the newspaper insulation. I use the same polyethylene that I use to make the plastic greenhouse (Photo 71). You can use four separate pieces fastened from one 2 by 3 inch to the next. Cut pieces of plastic 24 by 30 inches. The sides should be folded over a few times to make it stronger and stapled to

*71• Plastic covering stapled to 2 by 3 pieces to hold newspapers for insulation*

*72• Newspaper Box, fully insulated and being watered. Plastic has been pulled over the top and stapled to inside of box*

the 2 by 3 inch. It should also be stapled to the base. Leave enough room to put about one inch of newspapers inside, then staple it around the top to make a nice watertight job (Photo 72). After you do one, you will be able to make corrections.

## Hosing Plants

Many years ago when I started to use a spray of water to control insects, I first noticed that my plants looked healthier than my neighbor's. Although he did indeed get good results (I was his teacher), his plants did not sparkle and were curlier than mine. It was then that I realized that plants can possibly absorb water through their leaves, not only their roots.

Did you ever notice how fresh and healthy plants look after a rain? This same effect can be duplicated by hosing your plants, especially on very hot days. Many will disagree and, in fact, gardening books advise against letting water touch the leaves except when it rains. Surely my results are valid proof that I am doing the right thing. Such benefits, though not immediately apparent, can mean slightly larger tomatoes, which may also lead to an earlier harvest. And it's a boost in morale to see your plants sparkle.

I do not advocate wetting plants until the temperature is such that plants will benefit from a cool sprinkle. Did you ever feel like walking through a lawn sprinkler on a hot day? These are the days you should hose your plants. Of course, it isn't that hot in the morning when I go to work. But I won't be around in a few hours when that hot sun starts to beat down, so I hose my plants in the morning when I water them. First I water the Soil Bed. Then, using a medium spray, I wet the large lower leaves only, staying away as much as possible from the flowers or blossoms. Half a minute or less is usually sufficient.

I like to have my plants dry before nightfall. But if we are in a heat wave above 90 degrees, a cool sprinkling of water may be the only way to cool the plants enough to set new tomatoes. So when I come home from work on such a hot day I aim the hose high above the plants so the water will fall as "rain."

When I am home on weekends I go out about three o'clock in the afternoon. If it's another scorcher, I use the method just described, shooting the hose high above the plants to settle as rain. As with everything else, you must use a little common sense. Omit the hosing on cool or rainy days. You should start when the days are warm enough to go swimming, and cease operations about Labor Day. Remember, you never can do much harm even if you water on a cool day. After all, it does rain on cool days and no serious damage usually results. You will probably get some criticism from know-it-all gardeners. Tell them that scientists have found that the old way of never wetting plants is now obsolete!

## 35 • Blossom-end Rot

As the name implies, when Blossom-end Rot, or BER, is present, the bottom or blossom end of a tomato develops a black spot. It is important to know when you have such a problem so that you can remove the affected tomatoes in their early stages. You should start looking for these black spots when the tomato is about one inch in diameter —about the time needed for the disease to work its way to the bottom.

I sometimes use a mirror the way a dentist does. This allows me to easily see under each tomato without crouching down. If an afflicted tomato is left on the plant until

*73• Tomatoes afflicted with Blossom-end Rot*

maturity, you will be quite disappointed to find the inside usually completely decayed (Photo 73).

This rot is not actually caused by a communicable disease; rather by adverse growing conditions. If you try to look up the cause of BER in books on vegetable growing you will probably get some of the following:

BER is found more in staked plants than those allowed to sprawl on the ground.

Though not quite clear about the actual cause, we do know the plant does not take in water evenly.

Some plants are resistant to it.

BER is caused by a shortage of calcium, and lime should be applied to the soil, etc., etc.

I was not happy with what I read about BER, so I decided to make tests of my own. It is worse when there are many days of 90 degrees and above. Intense heat seems to be a contributing factor, causing some wilting. It can also occur if the plants wilt from standing in water for a long time, such as after flooding. But BER almost never appears in the cooler days preceding and following the hot summer months.

In my first experiments with a box unit I did not use any insulation. The box was made out of 3/4 inch wood, and when the hot sun beat down on the box it got hot enough to almost burn your hand. As I put my hand inside the box and against the wooden side I could feel the warmth that penetrated.

Usually a few weeks after a heat wave of 90 degrees and above, all the newly formed tomatoes developed a BER spot on them. However, all the tomatoes that had set in the cool days of May and early June did not have this problem. I now knew I had found the answer to my problems. When high temperatures dry out young roots exposed or in contact with extremely hot materials plants usually wilt a little. Specifically, heat from the walls of my box were drying out the newly-formed roots: These had a direct effect on the new tomatoes that were being formed up top. Here, then, is my theory:

This effect is similar to a stroke in humans. The person does not die, but develops a malfunction. In our case the small tomato is suddenly deprived of vital oxygen and other nutrients that the root system supplies. This sudden curtailment manifests itself by a rot spot which is not visible until it gets big enough to see, a few weeks later.

BER is usually not a problem when plants are allowed to sprawl on the ground, as it is with plants that are staked. Thus we can expect that BER may appear on your staked plants, but only during hot summer months. But it is usually not a serious problem, especially if you follow my directions.

### Remedy for BER

I have read in a number of gardening books that if BER is a problem, additional calcium or lime should be added to the soil. In my opinion such advice is not only bad, but it can cause really serious problems. Anyone who has studied soil and the growing of crops knows only too well the seriousness of making your soil too alkaline. Iron precipitates out, causing a very serious deficiency in the plant many times worse than Blossom-end Rot. There is no point in trying to help the situation by watering, since it all happened weeks ago. The answer is prevention.

The next year my box was well insulated from the sun. I actually made a box within a box, and filled the space between the two with newspaper. This reduced the problem about 95 percent, since my inner box remained cool at all times.

I still developed an occasional tomato with that black spot on the bottom, and I concentrated on finding the leak in the perfect setup. It was quite obvious that the only place where the sun could dry the roots was at the surface. While watering with a hose it is very easy to wash away the protective soil, leaving the roots exposed. Here, a good mulch is important (Chapter 26).

## 36 • Your First Red Tomato

Year after year I still get choked up at the first red tomato of the summer. All those weeks of waiting, watching, working, and worrying have paid off—you are about to taste your first vine-ripened tomato of the season!

It is not likely that you will have to wait until late August to get your first tomato if you used my method. If you didn't quite make that July 4 deadline, you should try to find out why and make corrections.

## Some Possible Answers

Having a good healthy plant about 10 to 12 inches high, ready to go out for hardening in late April, will set you on the right road. If your plants were not tall enough start a little earlier next year. If your seedlings were not great you can practice during the winter months starting in December. Of course, this is only *practice*. You shouldn't start any sooner than March 1 when growing the plants to be used outdoors. The use of the plastic greenhouse will help you get earlier tomatoes, as will the use of a fruit-set spray.

## Taste

Generally speaking, many things can affect the taste of a tomato: the conditions under which it is growing, the variety, and whether it is allowed to ripen on the vine, or picked prematurely and kept somewhere to ripen.

Different varieties will have slight flavor differences, but I have found that any plant making a rapid growth, especially during the height of the growing season, can't produce a bad tasting tomato.

Many people report that the small cherry tomatoes found in vegetable markets seem to be on the tart side. I am inclined to agree that these smaller ones, though great for salads, do not have that smooth, delicious taste that the large red ones have. But they are available much earlier.

I have found that a red tomato off the vine in early July, is superior to one picked in August. Tomatoes harvested later will also have tougher skins since the cooler nights and hot days seem to have this effect on them. However, even though harvested in September or October, the taste is far superior to tomatoes picked green, as in most vegetable markets.

How do you know when it's time to pick the tomato? This question has been asked by beginners, and it's not as funny as it might sound.

For many years my wife, Anne, would eat only a firm tomato, even if it was a little green on top and not that rich red color you see in seed catalogs. This was a little problem, of course, since I would wait until the tomato fully ripened on the vine. But with plenty of tomatoes, she picked and ate what she liked and I did the same.

One day a few years ago I was ready to eat my dinner, and was just about to rush out and pick a luscious red tomato. But I was a little tired, having had a hard day at the office, so I settled for one of my wife's firm ripe ones. "But it has a little green near the top, and it's not fully ripe, how can it taste good?" I said.

After some thirty years of eating and growing this delicacy, I now eat only the firm tomatoes with a little green near the top. This is somewhat embarrassing, since for years I have held that there is nothing like eating a vine-ripened tomato almost dripping with blood-red juice!

I suggest you try it—you might like it. If you know someone who won't eat tomatoes because of that distinctive tomatoey taste and smell, persuade him to try one. It is less juicy and it does not seem to have that typical odor and taste so common in tomato juice and sauces; instead, it almost tastes like fruit without a sweet taste.

## Do Not Confuse With Tomatoes Picked All Green

The tomatoes I speak of are almost completely ripe inside and should not be confused with those picked immaturely green and left in storage for weeks.

Indeed, most beginners are surprised to find that when tomatoes are ready to be picked they don't look like those in the store. Store tomatoes were probably picked green so they can be shipped and handled without damage. The smaller ones in boxes were probably ripened in drawers where they are kept cool over a long period. These tomatoes have a distinctive, even pink coloring inside and out.

On the vine a tomato starts to turn red from the bottom or blossom end first. The very last place to turn red is near the stem where it is held. If a tomato is subjected to heavy rainstorms in its late stages, it will draw in water so fast that the fruit will crack near the top. Cold nights and hot days, and other unusual conditions, will also crack the tomatoes. Obviously this can't happen to the tomatoes you buy in the store, since they were probably picked before the tomato ripened. It is in the last stage that cracking usually occurs. But if it is cracked, just slice off the top of the tomato, as I do.

## Don't Refrigerate

"You can't put a tomato in a refrigerator," I told a lady visitor. "It's almost like putting catsup on an expensive steak. You can taste the delicate flavor only at room temperature." Most times I will pick the tomato only a few minutes before eating; I can detect a tomato that was picked yesterday.

Winter tomatoes are a different story. I must admit that I am not as fussy about my tomatoes when I have to buy them. These are tossed in a salad, and once you add the dressing you can't tell the difference anyway. But I still refuse to allow them in the refrigerator. Once I pick my tomatoes I keep a few in the kitchen on the counter top. The rest I place on my bar downstairs.

The old idea that the sun will ripen a green tomato is

bad news. It's only common sense that if tomatoes are placed on a sunny windowsill, the hot rays will penetrate the inside—you'll end up with cooked tomatoes if they're left there long enough. A cool dark basement is best for storage of a week or so, also for those picked too early.

Finally, there is no point in leaving your red tomatoes on the vine—unless to impress some expected visitors. The quicker you get these ripened tomatoes off, the less strain on the plant so that it can better set new tomatoes and ripen the others.

## 37 • Controlling the Green Caterpillar (Tomato Hornworm)

Possibly the most feared insect in the vegetable garden is the green monster shown in Photo 80. Because of its presence among tomato vines, many beginners have been turned off. But this green cat, as I like to call him, isn't really a serious threat. Nor is he difficult to control —especially in our units.

The life cycle of this creature is most interesting, in that his parents and offspring do not resemble him. Actually, he is one part of a four-stage cycle.

This caterpillar or larva, as it is correctly called, comes out of a tiny egg laid on the leaf by the adult sphinx moth, which resembles a fat butterfly with grey wings. When full size, the caterpillar will find a burrow in the ground and turn into a pupa, remaining in this stage all winter and part of spring. When the weather warms, the pupa will one day split open and an adult moth will emerge. After mating, the female of the species will find the plant that nature

*74. Tiny holes in leaf indicate young tomato hornworms at work*

intended her to go to, and in the case of the tomato plant, lay its tiny eggs under the leaf.

This change from egg to caterpillar, pupa, and moth, and finally back to egg, is called metamorphosis. Meaning a dramatic change in physical appearance, from one stage of life to another.

The green caterpillar will not make an appearance in the garden much before July 1, and many years, even later, so it doesn't make sense to start worrying in May. The war against this potential enemy starts at the beginning of July.

Never try to look for the small caterpillars themselves, as they are almost impossible to see even with a trained eye. Instead, look for black droppings as shown in Photo 75, or a leaf that has been partially eaten (Photo 76). Examine each tomato leaf for tiny holes (Photo 74). On the underside of the leaf you will see these little green crawlers about 1/4 to 1/2 inch long. Removing them at this stage will save you a lot of trouble later. I squash these tiny caterpillars with my fingers.

75• *As the caterpillar grows, you will notice black droppings . . .*

76• *. . . and more heavily-chewed leaves*

77• *Recognize his silhouette*

78• *This photo shows how the hornworm holds onto bottom of leaf while eating*

If you didn't get him at that stage, in a day or so he will be about 1 inch long and will chomp away at the leaves as in Photo 77. He might be under that leaf if he didn't move to a new place—an old trick. However, insect control is much simpler using my neat, compact units. If you can't find him under the leaf, even though you see the droppings, crouch down and look up as you circle around the entire unit. You are very apt to spot him as he is possibly in transit.

It is important to recognize his silhouette, and I've provided a number of photographs to study. Photo 78 shows how he holds on and chomps away. In Photo 79 he was nice enough to show me his beautiful set of teeth.

Of course if he reaches his final stage, he will be the size of a man's finger and strip an entire branch of its leaves in a day. You can tell the size of this insect by the size of the black droppings he leaves. In his final stage, when he is most destructive, the droppings are about 1/4 inch in size

79• *Mature stage*

*80• Droppings are a fair indication of what size caterpillar you are looking for. Note tail where you can grab him*

(Photo 80). He is never at ease in our units since he does not like to be up high where man and bird can spot him. However, do not depend on birds except in places where there are few trees. They have plenty of food up high without taking the risk of being caught by a cat down below.

Little white objects on a caterpillar are the cocoons of a parasitic wasp that are slowly destroying him, usually in his late stage. Do not destroy these or the caterpillar; instead take him off the plant and toss him in the weeds somewhere, since these parasites are vital to the control of this insect.

I have never really found tomato hornworms to be a serious problem, and by September they will gradually disappear as they retreat underground to pupate. Note: I am not suggesting that you squash large caterpillars with your fingers. I grab them by the tail and destroy them by stepping on them on a hard surface. You can toss them in a pail of water or into a closed garbage can.

# 38 • Nematodes and Fusarium Wilt

In my many years of growing tomatoes, neither I nor anyone I know of has had any problems with nematodes. However, from what I've read, they are little worms that are supposed to produce swellings or knots on the roots —and of course create serious problems. As with all soil-based insects and diseases, once in the soil they are there to stay. They can only be removed or destroyed by fumigating or heating the soil to high temperatures in an oven before adding the plants.

There are alternatives. One is to use special seed resistant to this problem—it will be marked "N" after the seed in catalogues. The other answer to the problem is to use completely new soil; this is not too difficult to do using our method, since so little is used.

## Fusarium Wilt

One of the most feared tomato diseases is fusarium wilt. Actually there are two such diseases: Verticillium and fusarium wilt. Perhaps because the first is too hard to say, we usually speak of all such wilting plants as being affected with fusarium wilt.

In the good old days when I had to grow my tomatoes out of a box in some small backyard, there was not a year when I did not lose plants to fusarium wilt. Growing in the same soil year after year there was almost no way to avoid it. Luckily things are different today and an affected plant is the exception rather than the rule.

Once you pass into August it is possible to find symptoms of these diseases, but they are more common toward the end of summer. If you come home from work and see a wilting plant near healthy looking ones, this is a good indication that the plant has been affected. Naturally, you will try watering. The plant might appear to snap back, but if it wilts again the next day you can be sure that the plant will soon die. There is the possibility that this plant might have a borer slowly destroying it, but chances are it is fusarium wilt.

## No Cure Available for Sick Plants

Once the symptoms develop, there is little we can do to save the plant. This does not necessarily mean that all the plants in the same unit will be affected. There is no point in taking any action other than to sit it out. The tomatoes on the plant are not affected; the plant dies because its main feeding lines are closed off, and the tomatoes will be similar to those picked green, possibly a little tastier.

As with many diseases prevention is the best remedy. Next year take precautions. Here are a few things you can try.

1. Use different soil, or rotate your planting site.
2. If you have only one unit, during the winter months you can bake the soil in the oven if so inclined.
3. Use plants or seeds resistant to Verticillium and fusarium wilt.

One reason why I abandoned the old-time favorites is that many are not resistant to VF diseases. I also feel there is no point in using a hybrid if it's not disease-resistant.

But there's no such thing as 100 percent disease-resistant plants. Even after using them, you might still lose some. I suggest you just keep trying new disease-resistant seeds or plants; you might find some better than others.

# 7 Other Multi-Plant Strategies

## *39•Super Yields From Early Varieties*

A few years ago I posed for a photograph standing next to my tomato box. The resulting picture was so sensational that it was reproduced all over the world. I thought then that I had reached the limit of the Multi-Plant Method.

I now have another very unusual tomato picture; this one shows a myriad of smaller tomatoes, many about the size of a golf ball, all growing out of a 20 by 20 inch hole in the ground.

Photo 81 shows a count of over 200 tomatoes.

The reason for this profusion of tomatoes? Eight plants are growing where normally one would, providing tomatoes a few weeks before the main crop (about June 20). This unit is highly recommended wherever growing space

*81• An early variety ground unit, photographed from high on a ladder. White circle just above label is a golf ball. Note the Masonite sides of the basic unit. You can count over 200 tomatoes*

is available as the answer to a large early crop. For a full month and more, we have had enough of these smaller tomatoes to supply a few families. This unit could also be the answer for northern gardens faced with a very short season.

There are many seeds and plants to choose from, depending on the size tomato you want. I have also grown four different varieties in the same unit, and the results were quite interesting. One was 'Small Fry,' about 1 inch in size. Another was 'Spring Giant,' almost normal in size, but earlier than the main crop. Both plants were All-American Winners. The plants shown in my photograph are Harris's 'Presto.'

Burpees's small tomato is 'Pixie,' and you may also grow

eight Italian-style tomato plants in one unit. A very popular plant seems to be 'Roma VF.' You can start ten plants on a window sill with two for spares, just in case.

## Cultural Instructions

In actual practice, you would prepare the units exactly the same as for the regular crop discussed earlier. But in order to get the earliest possible harvest, you would have to set well-established plants into your plastic greenhouse, according to schedule. Even if you did not use the protection of the greenhouse, you would still have good results, but your first red tomato would not be ready so soon. In this case, you would set plants out later. (See schedule without greenhouse.)

I have never tried these plants in a box. While results would be good, you might have problems trying to spread out the plants as I do on the ground. But even if you allowed the plants to hang over the sides of the box, you should get a good crop.

When setting plants into the Soil Bed, you must be especially careful because you will be placing eight plants in a small 20 by 20 inch area. With the plastic sides in place, it is somewhat of a challenge. But I have done it a few seasons and had no trouble at all. You might add the plastic sides after plants are in place, but be careful not to damage plants.

First, place one plant in each corner a few inches in. Then place one plant on each side, between the two corner ones. (Follow the instructions given earlier for setting out plants.)

Once the plants start to spread (possibly as early as one week, if well-established plants were set out) you should use 2-foot bamboo sticks to train them outward. It is important to get each plant up and out as soon as possible.

Soon you must enlarge the size of the plastic greenhouse to about 2 feet by 2 feet. (This is not necessary when using the normal four plants.)

When the greenhouse is finally removed, fix eight 4-foot bamboo sticks outside of the 20 by 20 inch square. This becomes your permanent setup.

## Pruning

These plants are very difficult to prune to single vines. But you should try, or you will have a jungle. (See Chapter 18 on pruning to a single vine.) If nothing else, you should surely remove the suckers down below. Look down into the Soil Bed; if you see another vine or sucker starting to grow, no matter how big it is, break it off. Fruit set solution is not needed with these small size tomatoes. They set well by themselves.

It is important to add liquid fertilizer once each week, starting when the tomatoes are 3/4 of an inch to 1 inch in size. Start with half concentration for a week or two. Following the instructions in Chapter 30, half fill a gallon jug with concentrate, and add water to fill the rest of the way. Once each week first water well, allow to drain about fifteen minutes and then slowly add the liquid solution, spreading it all around. Since you will now be in the warm weather, water every day, using 3 or 4 gallons. Make sure your mulch is in place so that you will not dig up the soil.

When tomatoes start to turn slightly yellow, increase the fertilizer to a full concentration, as for normal feeding. Withholding plant food at this time will delay the ripening process. Hose the plants only if tiny insects are present. Carefully use a fine spray on them.

One day you will go out and find about one third of the tomatoes turning red!

Finally, don't become too concerned if some of the

plants are starting to dry and die out. This is almost a natural way of life for these early varieties. The plants do not get too big, but usually deteriorate once their heavy load is harvested, although there are some plants that make it through the season. After most of the plants are nearly dead, consider a second planting of other crops in these units.

## 40 • The Weekend Gardener

I have had many requests from people who have summer homes and only go there on weekends. Since they cannot water every day, how can they use my Multi-Plant Method? I have tested a unit for just such gardeners, and although not quite as good as my regular units, results were very satisfactory. This is a ground unit and it will be described in detail in the following pages.

Of course, there is no substitute for daily care, but this unit was only watered Saturday and Sunday and did quite well.

The Soil Bed was 24 by 24 inches instead of 20 by 20 inches as in the standard ground unit. I prepared the soil as usual, except that I added extra organic material. I used pieces of 1/8 inch Masonite to form a boxlike top as in regular units, but these pieces were buried only 12 inches deep. I used four pieces 24 by 12 inch, so that about 6 to 8 inches went below ground. (It can work well without Masonite.)

After I set out my plants as per the schedule, I set up the plastic sides with no covering on top as wind protection and pruned my plants to a single vine. Many changes were made in the cultural techniques, especially in watering and feeding. When the warmer weather set in, about late May,

I removed the plastic sides. Rather than use pieces of bark for mulch which might not be available in country places, I prepared dried weeds and hay and set about six inches of this mulch into the Soil Bed about early June. I watered every week—a little during cool days, and much more during the summer months.

When the green tomatoes were about 1-1/2 inches in size I started feeding, using a one-half concentrate every week. A regular gallon of concentrate used weekly on my regular plants was put into two gallon containers; then water added to make a full gallon.

Friday night or Saturday morning, I watered well—about three separate waterings of almost a minute (this was during midsummer). Sunday morning I watered again two separate times of about one minute each. I waited about fifteen minutes or so, then poured on the gallon bottle of the weaker liquid fertilizer; this was repeated every weekend. If the mulch looked matted down, I added a little more. I also hosed my plants on hot days.

## Possible Problems

Insect control appears to be one area where you could expect trouble. But even without adding insecticides, I did not have much trouble.

When I saw some tiny aphids in June, I used a fine spray, as in the book. I repeated the same the following week and soon they left. I did become concerned when I saw a few tiny green caterpillars. Late Sunday, as though I was leaving for home, I carefully looked for all the small ones I could find. The next weekend I looked again and found a few growing quite large. Except for a few half-eaten leaves, not too much damage was done. However, I am sure if left alone another week the creatures could cause some damage.

The plants did very well. They were just as tall as in my regular units, but had slightly fewer tomatoes on the vines. I had red tomatoes not too long after the regular units. By midsummer and early August I was picking many 3/4 pounders. Visitors looking at this unit thought it was absolutely great. I had to explain it was not one of my better setups, but only a test.

Plenty of organic material in the Soil Bed and a thick mulch are very important. Setting the plants out according to schedule and using the plastic sides got the plants off to a good start. Of course a good disease-resistant hybrid is recommended. In September and October, the unit was at its best, really loaded and healthy thanks to the cooler weather.

## 41 • The Unit for Infrequent Visits

The weekend gardener uses a larger Soil Bed with extra organic material and a good mulch. But if you don't think you can visit the unit every week, then I would suggest something else—a large bed of rich soil. Set in four plants and do not stake them.

I am suggesting a modified version of the way many farmers still grow their tomatoes which would take too much room in a small garden. They prepare the ground to a depth of almost 2 feet by turning under rye grass, manure, etc. They place their plants about 4 to 5 feet apart and almost forget about them. The plants, sprawled on the ground, form their own mulch. I still recommend four plants in the same hole because of the many advantages. But there will be modifications to the farmers' way.

If you are lucky enough to be near a wooded area, you

can get some well decayed material to be incorporated in the soil. As a matter of fact you may even have so-called "virgin soil" with much organic material in it.

But I will be very frank and say that unless you can find pretty good soil that will retain water for long periods, it might be unwise to waste time. I can't see how a sandy soil, even with organic material added, could retain moisture for two or three weeks during the hot summer months.

Assuming that you found some good country soil in a site with plenty of sun, you should prepare a section about 3 by 3 feet. This will mean digging it down to a depth of about 18 inches.

If you follow directions and use good hybrid plants, one unit can supply all the tomatoes one family can eat although they will be a little late getting started. Add some organic material, some 5-10-5, and lime.

## Detailed Instructions

It would be best to dig up the earth from this 3 foot by 3 foot section and put it in a pile, if for no other reason than to loosen up the soil. If you don't feel up to it, dig it up as best you can and as deep as you can. When you dig out the hole, try to keep all the good topsoil separate. Look around the area for decayed material and place it on the pile. If there's none available, use peat moss. Spread out the soil, mix the organic material with it, add some 5-10-5 and lime. If you have much rich, black soil, go easy with the 5-10-5, although a light sprinkle can't do much harm.

If you don't dig up your entire Soil Bed, try to mix the organic material plus 5-10-5 and lime, as best and as deeply as possible. Mix it all well and dump it back into the hole *(Masonite sides should not be used)*. Here again, allow a small depression to form to allow water to collect during a rain. Water well before adding plants.

The four plants should be set in the center, 12 to 18 inches apart. Observe all the directions about setting out plants, and avoid cutworms by using a guard around each plant. Once your plants are set out, you can make a wind guard around them by using four sticks and clear plastic or old rags and burlap. Remove it once the danger period is past.

Don't stake the plants. Allow them to lie down naturally. Each plant must grow outwards, occupying about 90 degrees of a large circle.

All the lower suckers should be removed to allow the four plants to work their way out. This will also encourage early tomatoes to set. Once the plants are out and away, and after your first flower clusters have set tomatoes, you can allow more of the vines to grow. In other words, do not remove any more suckers once the plants have room.

## Watering and Feeding

On cool days don't water at all unless it did not rain for many days. But a little can't hurt before you leave. At the time to feed the plants (when tomatoes are 1-1/2 inches), soak the ground well, allow to drain and add a gallon or two of the one-half concentrate described before. Each time you visit your garden water well and add the one-half gallon concentrate, spreading it around evenly. Add two gallons because of the 3 by 3 foot area.

When the hot weather starts, lay down a 5 to 6 inch thick blanket of hay or similar material over the entire 3 by 3 foot section. You should also prepare more of this dried grass or hay to allow the plants to lie on it.

## Insect Control

Unless you want to flood the plants with insecticides, you might try growing marigolds around the plants. I have read that this has an effect on the insect population because of marigold's unusual, pungent smell. Of course when you are there keep an eye open for the green monsters and pick them off.

## Growing Near the Beach

Finally, some tips for our lucky seaside friends. If you have people living there all summer, without a doubt your best bet is the Newspaper Box.

Otherwise, follow the directions for the weekend gardener.

If you are living at the beach every day, there's no reason why a regular 20 by 20 inch ground unit can't be used. Excavate the sand and bury the side Masonite pieces the full 20 inches down. Use topsoil and peat moss as soil, and follow directions in the book.

Your main problem, of course, is getting an early start. This may be difficult since most people don't go to the beach until late June. However, very little water is used in the cool days of May and early June. Once a week is enough even for the box. Plastic sides will help keep the cool winds off the plants until they get started.

I have been told the salty air has no ill effect on the plants, but hosing them during hot days is recommended to freshen them and eliminate any build up of salt spray.

## Automatic Watering System

If you are growing a box unit on a balcony or elsewhere where it must be left for the weekend, your first choice would of course be to have someone water your plants while you are away. However, if this is not possible, you can install an automatic watering system—without a hose!

Hopefully you have an electric outlet. Use a large plastic garbage pail filled with water. Then get a small water pump and a time clock to dispense about 4 gallons a day (two in the morning, two in the afternoon). You might shield the plastic pail so the water will not get too hot.

If it works, write to me in care of the publisher.

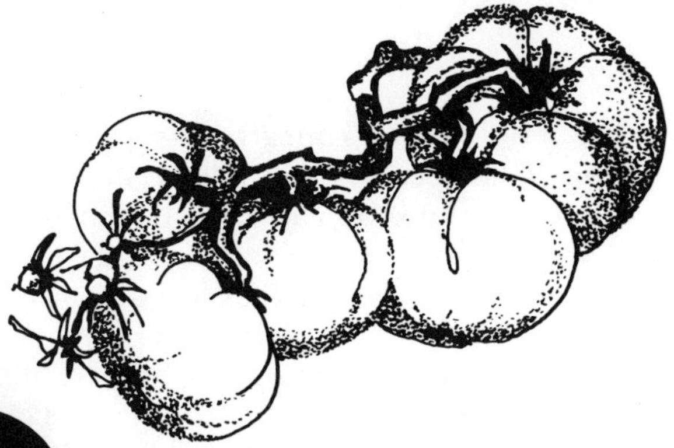

# 8 After the Midsummer Harvest

## *42 • New Tomatoes From Old Flowers*

Regardless of weather conditions, the same procedures such as watering, feeding, pruning, tying, and looking for green caterpillars, are continued throughout the season.

If you followed directions, come late August your plants will have reached the top of their stakes and be still healthy looking, except for some brown leaves down below. It is a natural condition for the lower leaves to turn brown and fall off. You will notice that after many weeks of lush harvest, the plants will have slowed down a little. I refer to this period as "getting their second wind." The cooler nights have a little effect, but the real cause of this slowdown is past heat waves during which no new tomatoes were set. It would normally be time to harvest those clusters of tomatoes, but they do not exist.

But don't be downhearted—there are good days ahead! This cooler period, especially at night, is ideal for setting tomatoes, and soon you will see your plants loaded again. Because of the cooler weather there will be many new flowers and small tomatoes forming near the growing tip. But if you look at the lower part of the vine you will notice small green tomatoes popping up here and there that suddenly seem to come from nowhere. Many gardeners are not even aware of what's happening. If you think about it, you will realize that all the tomatoes were already picked off so there must be some reason for this bonus crop. There is, in fact, a very interesting explanation for this strange phenomenon. I call it a "delayed action" from past flowers or blossoms that did not set.

Under normal conditions, new tomatoes keep setting up top and a steady supply is usually maintained. However, a week of temperatures in the nineties, especially with a load of tomatoes on board, can leave you with a dried flower cluster and no small tomatoes.

It's at times like these that Mother Nature seems to compensate. Some of those flowers that did not set a tomato early in the season will surprise you later on. Normally, when the flower dries out you either get a tomato or the flower falls off. But there are some blossoms that can't seem to make up their minds. They just freeze—that is, they don't swell up and they don't fall off. Many will stay in this dried out state for quite a while and one day will do a sort of second take and produce a tomato.

You can't expect to get a tomato if none are there. You must see a small one sitting in a dried out flower. Look at some of the old lower clusters that did not set well and you'll see what I mean. Look to see if the green leaflets are still in place and still look alive. These are called sepals or, collectively, the calyx. These green things will be what is removed from the tomato after it is taken from the plant. Inside will be the tiny tomato.

If these leaflets are still green, your chances are good. Keep an eye on them, and one day you may suddenly see a tomato emerge.

This bonus crop, as I call it, usually will start to appear in late August or early September, after the heavy crop has been picked. Usually this won't happen until after the plants have given up a heavy yield. Of course the plant must be in good shape with plenty of leaves. My theory of what happens is based on common sense. The strain has suddenly eased and the plant is looking for new tomatoes to fatten up. With most of the strain gone, this large healthy plant can go back and get the stragglers, so to speak.

You will see these tiny tomatoes usually starting to get bigger as the late season approaches. Once they start to grow they will usually reach maturity in a very short time. I have seen what appeared to be a little white dot blow up into a one pounder in a few weeks.

## 43 • Bamboo Extensions on Stakes

One reason why your plants get so tall is that they are pruned to a single vine, instead of being allowed to branch out. This way of pruning has many advantages, but also one disadvantage—the plant grows extra tall. By late August, according to my schedule, the plants will reach the top of their tall permanent stakes and say "What do we do now?"

Some writers recommend topping them or cutting off the growing tip. I also recommend topping, but about the beginning of October, not September. Topping will allow the green tomatoes on the plants to ripen faster.

Under some circumstances I may even recommend this

earlier. If you are growing plants on an apartment balcony you can't let them grow higher than the balcony above. I recommend that you cut off everything up top in order to allow side shoots to grow and possibly set small tomatoes. So, as you approach the top, allow a few suckers to grow and hang down. If you're lucky you might get some tomatoes on these vines.

Of course if you are where the sky is the limit, it is much too early to retire your units by the end of August. Come Labor Day, you have almost two more months of harvest. So it is important to let the plants continue on their upward climb. I do this by adding 6-foot bamboo sticks, purchased at most garden centers, to the stakes. Thinner and less expensive four-foot or shorter bamboo sticks may also be used. Most of my six-foot sticks are many years old. In the early days I didn't cut off the tops of my plants come October, and the plants got somewhat taller. Now that I top the plants to give the smaller green tomatoes the extra energy, I think cheaper, four-footers would be adequate. Different varieties of plants and growing conditions in other parts of the country would dictate the proper length to use.

*82• How bamboo extensions are fastened to existing stakes*

*83• Vines climbing up extensions. Note new tomatoes setting above original stakes*

The bamboo sticks must overlap the stakes by about one foot and are held to them with wire at three different places (Photo 82). Have ready a roll of steel wire plus cutting pliers and regular pliers. It's helpful to have an assistant hold the stakes together while you tie them, but if you don't have one, hold bamboo and stake together with masking tape. Cut off pieces of wire about 8 inches long. Wrap them twice around both pole and stake, then twist them; tighten them using the regular pliers, but do not twist too tight or the wire will break. Now the vines can continue their upward climb (Photo 83).

## 44 • First Frost

All good things usually do not last. Therefore, unless you live in the warmer climates, your tomato plants will one day be destroyed by freezing weather. A temperature of a few degrees below freezing can shrivel up your plants in a few hours.

Plants will probably still have many large red and green tomatoes on them, and you must try to salvage these before they are ruined by the oncoming freeze.

But when should you start picking them? It's like telling your broker to sell your stock when it hits the top! If the tomatoes are removed too soon, you'll miss out on weeks of good growing time, but too late means disaster.

After many years I have become quite experienced and can tell almost exactly when I should remove my tomatoes. I have also found that a premature or light frost will strike first, leaving many weeks of good growing weather before a killing freeze sets in.

It is important to know how to deal with this first frost. I use a window thermometer, weather forecast, and plenty of common sense and experience to assist me. Once September arrives I put myself on the alert. From then on there is the possibility that an unusual early frost can strike. This is not information from the weather bureau, but my own findings. In my area it will probably not be until early October, but the exact date varies from year to year.

Since it is important to know when the first frost (thirty-two degree temperature) will hit your area, listen to the weather forecast and keep an eye on your outdoor thermometer, especially about 11 o'clock at night. Once you start to get readings of 40 degrees and below, it is an indication that time is getting close. A reading of 40 degrees should not cause panic, but it might be wise to remove tomatoes that have started turning red. A reading of 35 degrees at midnight could at worst mean a light frost which probably will not do any damage, but it's a good time to remove all the large green tomatoes to ripen indoors. This leaves tomatoes that are about two inches across or slightly larger. If it is still early, you should risk these smaller tomatoes. If you pass the crisis, in a few weeks they could become much larger. But if it is already late October or early November and your readings are consistently cold, 32 to 35 degrees at night, then it is time to call it quits and remove all the tomatoes on the vine. They can be used to make sour tomatoes, or they can be left to ripen in a protected place.

People have asked me "What can I do to protect my plants from a frost?" If it is late, there is no sense to waste your time. But if it is an early frost, you can use a fine spray of water on the plants all night, since running water will usually not freeze. Besides, the temperature of the water in the hose is probably between 40 and 50 degrees, which will feel like warm water to the plants.

## Processing Green Tomatoes

If things worked out and you managed to cope with the oncoming freezing weather, you should have many green tomatoes of different sizes. These should be sorted out as to large, small, and bad.

The large ones should be washed and dried and kept in a cool place, possibly a basement, to ripen (but not a sunny window, as mentioned earlier, that can cook your tomatoes). Your basement is dark enough and tomatoes need not be wrapped in paper.

You should decide whether to use the smaller ones to make sour tomatoes, or take a chance on ripening them. Immature tomatoes and those you wish to keep for later should first be washed in hot water, then cold, to get rid of some of the bacteria that can cause decay later. Dry them well and wrap in paper toweling. Old-timers used newspapers because they did not cost anything. I still read where writers recommend wrapping tomatoes in newsprint. Don't these people know that printer's ink can be poisonous? The purpose of the paper is to keep the tomatoes from touching and absorbing moisture—not so much for darkness as most people think. Once wrapped in paper toweling, the tomatoes should be placed loosely (not squeezed!) in a brown paper bag.

Smaller green tomatoes must ripen slowly and be kept in a cool place about 40 degrees. Though fine for the large tomatoes, the basement is not cool enough. A garage or shed is an ideal place for them, but they must not be allowed to freeze. You should check the tomatoes every week or so to see if some are getting ripe, and remove any that start to decay. If you keep them cold, it's possible to get red tomatoes well into December—but you will miss those delicious vine-ripened ones you picked in July and August!

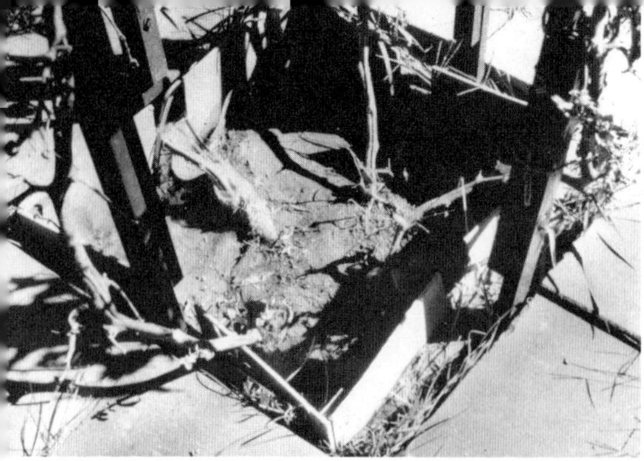

*84• Ground unit after a hard freeze, with bark mulch removed*

As the weather gets colder, place them in an extra brown bag and possibly cover them. As I said, they must not freeze. If no cold storage is available, yes—keep them in the refrigerator.

## 45 • End of Season

Once a killing freeze has reduced your plants to a dark green, wilting mass, it's time to clean up and prepare for next season (Photo 84). Your first step is to cut down the plants, but *do not* put them on the compost pile. Weeds, tree leaves, and similar material can be buried to make compost, but most vegetable plants should be taken off the premises. You have enough insects and disease in the garden without encouraging more trouble! I suggest you put them in the garbage.

Your next step is to remove the stakes from the box right after you've cut down and removed the old plants. Be careful not to lose the nuts and screws. Put the nuts and washers back on the screws, ready for next year. Your two-piece field stakes can be left together. The stakes should be hosed down and all strings removed. Dry the stakes and tie together. They can then either be stored in the garage or left outdoors all winter. A plastic covering will make a neat package. Pieces of clear plastic bag from the cleaner's are fine, and if the stakes are painted every few years they will last almost forever.

It is important to salvage the pieces of bark mulch at the end of the season. Remove it all carefully and bag it up.

Since I grew many vegetables, I also try to rotate the bark mulch next year if possible. Of course the only problem is trying to remember which is which! Since my experimenting with other vegetables is fairly recent, I am trying to work out a system.

## What to Do With the Old Soil Bed

If you are using a box, it must be emptied out before the winter sets in. If the soil is never removed, the box will last only a year or two before it starts to deteriorate inside. There is no rush to empty the unit. As a matter of fact, I prefer that you wait and try to wash out some of the fertilizer first.

You should stop feeding the plants possibly by early October, but keep watering at least a little while the plants are still alive and have tomatoes on the vines. Even after the plants have been frozen, I prefer to allow the rain to wash out any large accumulation of nutrients in the Soil Bed as a result of the weekly feedings. If you are on a balcony where no rain hits the unit, it might be wise to add water for a few weeks, then allow it to dry out a little. You have until late December to remove the soil. I am happy to report that I see little earthworms in the soil when I empty my boxes, even though much of the compost is disintegrated into humus and very little food is left for them. The worms tell me that my 5-10-5 is not poisoning the soil as many report.

I usually dump my soil on a pile, but there is no reason

why you couldn't put it in neat boxes or plastic bags. I prefer the open boxes so the soil can better be exposed to the winter weather. Make sure the boxes drain so water will not build up.

Once the boxes are emptied out and washed, you can possibly put the soil back in plastic bags and cover the unit. I have not found this necessary, but it might make sense on a balcony where space is limited.

## Excess Soil

According to my directions, you should add compost or peat moss to the soil each year. Actually you will still get good results next season even if you used the soil as is. As a matter of fact you can grow in pure sand as long as you feed and water the unit properly. The purpose of the compost is to help the water last much longer, and so that less is needed. Moreover, the organic material breaks down and enriches the soil, adding little goodies to it that are needed for good growth. But what about excess soil? Obviously you should start another unit possibly for other vegetables. You can also use this soil for house plants.

## Old Soil Bed

As for my ground units, I dig out the holes and fill them with leaves or scrapings from the lawn. I add soil to the leaves in layers and finally cover the top with more soil, leaving a depression for water to settle in.

Depending on how I feel and how much time I have to work outdoors (I have about 15 units), I usually also remove the pieces of Masonite carefully and allow them to dry out above ground. This of course is optional, but helps Masonite last much longer. I place them on piles to keep them flat.

## Newspaper Insulation

There is no point to remove the plastic and newspaper insulation from the Newspaper Box. It will protect the outside of the box during the winter. As a matter of fact, you can leave it on for another year. Possibly you might add new plastic over the old if it needs it. But unless you are opposed to its appearance, there is no cultural reason for removing the insulation.

If you are using the Deluxe Box, you need not remove the newspapers any more as long as it is protected from the rain with strips of Masonite or something similar.

# 46 • Growing Tomatoes Indoors Near a Window

A number of years ago, having had such great success with my outdoor plants, I decided to try my hand at indoor tomato growing. It was indeed a thrill to see these almost artificial-looking red tomatoes growing in my dining room!

Tomatoes never get the size they do outdoors. But the best news is that the soil mixture, and the watering and feeding schedules are almost the same.

Because I had to do my indoor growing from September to March or April, I was able to experiment for only a few years before my wife put a stop to it. My regular outdoor experiments kept me tied down from March until November, and it was only fair that there should be some period during the year when I could leave my house on vacation for a week or so. But in the few years that I did grow indoors, I learned many interesting things.

For example, I don't recommend you try to grow tomatoes indoors during the summer. The sun is too high in the sky to shine into your window far enough to benefit your plants. However, if you have a setup where you do get enough sunlight, there should be no problem growing vines in summer if it's not too hot.

By September, the sun will start shining in your window as it drops lower in the sky and it will continue shining at an angle until May. Midwinter is actually the time your room will be flooded with most sunlight, although not for as many hours as in summer. A southern exposure is best.

## Choosing a Seed or Plant

Regular midseason tomato plants that you use outdoors are not satisfactory indoors—the larger plants will produce large tomatoes, but the plants soon get too big. The only reason for growing the so-called mini or midget size tomatoes is that the plants don't get too tall. You must consider that your window is only so large, and once the plants are without sun you're in trouble.

Although I have used a midseason plant, and found it grew too tall, you can choose almost any of the many smaller varieties available in seed catalogs and garden centers. But you might try 'Spring Giant,' an All-American Winner. Outdoors, this fabulous plant never grows much taller than about two feet, even if you try to prune it to a single vine, which is quite difficult. But it can produce a load of large-sized tomatoes just a little smaller than the real biggies.

One of the reasons I don't grow it much outdoors anymore is that it is not ideal for the regular units and it is not quite early enough for my early-variety tomatoes. But it's still one of the greatest plants I ever saw, producing red tomatoes a week or more before the regular midseason

*85• Single vine grown indoors near a bedroom window*

varieties. It's disadvantage is that it does not do too well once the big crop has been harvested. But it is truly a spring giant, and I feel it would do well indoors.

## Getting Started Indoors

As mentioned earlier, a south window is ideal and before I forget, I should mention that you must remove the window screen. You should get as close to the window as possible, but the plant should not touch the cold glass pane. Though storm windows hold back a little light, I didn't remove mine when I grew indoors. A radiator can be a problem. You must shield the plant from its high temperatures.

You should set up a sturdy table or something similar. Only the top of the pot need be at window level. This allows the plant to grow taller and still stay in the sun. One of the disadvantages of growing near a window—you get no back light as you do outdoors. Therefore while one part of the plant gets strong sunlight, the back is almost in total darkness. Rotating the plant once a week will help solve this.

## Choosing a Container

I used a ceramic crock or pot about 9 inches high and about 8 inches across, with a hole in the bottom. But there is no reason why you could not make a miniature model of my newspaper box. There's no need to add the newspaper insulation, since the sun doesn't get too hot in winter. This suggestion is only for those who are handy with tools, although you can get the necessary materials pre-cut and assembling it is fairly simple. A box about 9 inches square by 10 inches high would be ideal. You can try one or two mini-sized plants in it.

## Preparing the Soil Bed

The soil for this unit should be the same as recommended for the outdoor units. But for anyone purchasing material, I would suggest buying half peat moss and half topsoil. Many readers have used this with good results. If you have access to materials in your yard, some good soil with compost mixed half and half is fine.

You will also need a small bag of 5-10-5 and lime. A 5 to 10 pound bag of fertilizer is sufficient and perhaps you can borrow a little lime, since you will need only a cupful.

Prepare the soil the same as for seedlings, working on a table covered with newspapers if you are indoors. Fill the container to the top, give it a good soaking with water, and allow to stand a few days to settle. Adjust the surface of the soil so that after being wetted down it settles to about 2-1/2 inches below the top.

Ideally, you should start in September with a nice stocky plant growing outdoors in a small container. But if you can't get a started plant, you must start your own from seed, possibly in midsummer. Plant a few seeds about an inch apart in the soil about 1/4 inch deep.

For directions on how to grow seedlings, use the information given earlier.

## Cultural Procedure

As with all my plants, you will need a good mulch. Medium or small size bark (not pine) works best. About 1 inch thick is deep enough; you aren't concerned with keeping the hot sun off in January. The purpose of the mulch is to prevent the top from caking or drying out, and you can pour water around the roots without it digging up the soil. Peat moss or other dried material can also serve as a mulch if bark is unobtainable.

Until the plant gets its first flowers, watering well once a week is usually sufficient. I suggest you rotate your plant at least once a week to give the back side some sun. I have even thought of using a rotating lazy Susan to make things easy. In any case, you'll need something underneath to catch excess water.

Most of these mini-type plants are difficult to prune to a single vine. But do pick off the first suckers. This will allow the plant to set earlier tomatoes.

When the flower opens it should be pollinated with a small artist's brush. Just poke it around inside the flower. The chapter on pollination will explain why. This is more effective than shaking the plants. Tapping the plant may work, but is not as sure as with a brush.

Once the tomatoes set and the plants get larger, I recommend daily watering and weekly feeding. Each day, pour in some water until it forms a little pool, then stop. About one quart or a little less, is right. As the plant gets big and stocky, you can add a little more. Once each week you should feed the plant with the liquid fertilizer described earlier. A gallon concentrate will last for many

*86• Close-up of bamboo stake. You don't have to worry about wind, so masking tape, as shown here, is okay*

weeks. First water, allow to drain about a half hour, then pour on the liquid mix—possibly a quart or less. Continue the same procedure throughout the season. You will also have to tie up your plant occasionally, using bamboo sticks.

At best, growing indoors near a window is only a novelty, and you should not believe ridiculous articles that suggest that growing such plants indoors or on a sill can help fight inflation. But it is a great conversation piece for you and your friends, and it was truly one of my greatest lifetime thrills when I ate my first tomato grown near my dining room window!

# Conclusion: The Multi-Plant Method for Other Vegetables

There are those who still look upon my new method as a gimmick which will soon be forgotten. I've received too many good reports to support the idea that the Multi-Plant Method will one day fade away.

When Prentice-Hall asked me to revise and expand this book, I told them that the Multi-Plant Method is no longer only for growing tomatoes. Besides my tomato experiments, which will continue as long as I am able, I have moved on to other vegetables and now grow many other things.

It is my belief that any vegetable plant or ornamental, except those such as radishes and carrots that do best growing alone or in rows, will do much better growing together in a deep rich cube of soil given my special watering and feeding. Moreoever, they will use less space, water, and labor per yield.

I have used this Multi-Plant technique successfully with

many vegetables. However, since I did not have enough detailed information available for each vegetable, it was decided that I would hold off and not include them in this book while I continue my tests and observations.

But "Keep snapping plenty of pictures," my publisher told me, "We can't sell books without pictures; readers won't believe you."

I must admit that I am very excited with the results from growing many different vegetables. As a matter of fact, someone called saying, "Your method works great on tomatoes, but I've also been using it very successfully with dahlias and gladioli."

Meanwhile, I hope that you experiment on your own with peppers, eggplants, cucumbers, pole beans, and all the vine crops.

## Preview of My Multi-Plant Results With Other Vegetables

For string bean lovers, you have to see it to believe it. The "Perpetual Bean Stalk" is absolutely unlike anything you have ever seen or read about—another one of my great mechanical creations. A maze of bamboo sticks about the size of a telephone booth, or a little larger on top, but easy to construct, becomes a permanent, neat unit that has withstood light hurricanes. Once set up, I never have to touch it all season, except to add a little water each day and feed it once a week, as with tomatoes. Starting about mid-July and continuing almost until frost, I pick garden-fresh beans the size I want and without poison, totaling an estimated 2500, from a single unit growing out of a 20 by 20 inch Soil Bed!

Then there are box and ground units filled with green peppers. Once the season starts, they seem to replenish themselves as fast as I pick them. Four healthy eggplants, producing all a family needs, also works in a box.

Would you believe that a cucumber box unit seems to do better than the ground unit? Let them sprawl on the ground and pick more than a family can eat, starting in late June. Almost any vine crop will work—cantaloupes and all squash. But you must try zucchini which is absolutely simple and produced a one-pounder about every second day starting in mid-June.

In a few months I will have the results from growing broccoli, cauliflower, and potatoes, again four plants to a unit. These last are growing in a spot that doesn't get too much sun and are close to bushes, but they all look great. My bush bean unit in a large square is also highly successful.

Now, as I work with other vegetables, I am finding more new ideas. I am sure you will hear about them soon. I will be spending all summer gathering more detailed information and plan to have a new book in print in the near future.

Happy gardening!

# Index